JORDAN CUCINEA

ONE MAN'S OWL

One Man's Owl

BY BERND HEINRICH

With Drawings and Photographs by the Author

PRINCETON UNIVERSITY PRESS

Copyright © 1987 by Princeton University Press
Published by Princeton University Press, 41 William Street,
Princeton, New Jersey 08540
In the United Kingdom: Princeton University Press,
Guildford, Surrey

ALL RIGHTS RESERVED
Library of Congress Cataloging in Publication Data will be found
on the last printed page of this book

ISBN 0-691-08470-X

This book has been composed in Linotron Trump

Clothbound editions of Princeton University Press books are
printed on acid-free paper, and binding materials are chosen
for strength and durability. Paperbacks, although
satisfactory for personal collections, are not usually
suitable for library rebinding

Printed in the United States of America by
Princeton University Press,
Princeton, New Jersey

To Stuart, who was braver than he knew;
to Bunny, who was a good provider;
to Thor and Theo, who didn't give a hoot;
to Margaret, who (for awhile) put up
with us all; and
to Erica, who always had a smile

CONTENTS

PREFACE

Bubo, a great horned owl, came into my life by accident, and I naively presumed we would soon part company. But he became a special owl to me, and to him I may have become a special person. In any case, he refused to leave me for a long time. We were intimately associated for most of three summers, as he shared my camp in Maine and lived in the surrounding woods. Eventually he became, like all free owls, a hunter.

If his weaning process was slow it was probably my fault. I felt inhibited from providing him with large live prey such as rabbits, squirrels, and other wild animals that he needed to kill to sharpen his hunting skills. My rationality for what was right for him more or less lost out in the conflict with my feeling for his prey, and as a consequence he lived primarily on road kills that I collected for him. But in attempting to salvage some justification for my emotional indulgence, I set out to learn as much as I could about his behavior throughout his maturation process.

Bubo's behavior fascinated me, and I kept a detailed diary of what I saw to preserve it fresh in my mind. The chronological flow of most of my observations of his development is preserved here in its original diary form, but I sometimes have added additional information to provide context. Also, the first appendix gives an entrée into the voluminous owl literature, as well as some general information on the great horned owl.

During Bubo's weaning to independence, I recorded most of his interactions with potential prey. Eventually I focused on behavior that I thought might have bearing on prey recognition and capture—behavior that was most relevant to his freedom. I hope that the reader will not be offended by my descriptions of his predatory behavior. I recorded it with the knowledge that nothing would have been changed by looking the other way, except that nothing could have been learned. In particular, I wanted to learn why some birds mob owls so vehemently. I thought that his interactions with prey might provide clues, and I summarize these thoughts, along with current scientific theories on mobbing behavior, in the second appendix.

I cannot explain everything I saw. There are often several possible interpretations of particular events. At the risk of providing perhaps too many and sometimes repetitive details, I give as much primary information as possible so that the reader can make independent judgments. On the other hand, for economy of space I present only those interpretations of the owl's behavior that seem most plausible to me. However, I include a bibliography at the end of the book so that the reader can explore points of interest in greater detail than is possible here.

This book is not meant to be a scientific treatise. Rather, it is a close personal look at a wild animal under seminatural conditions. Certainly, it would have been scientifically far more valuable to have similar observations of an untamed great horned owl from the wild. But a shy wild owl that lives in the dense forest is next to impossible to track from day to day, from nestling to adult. It is even more difficult to observe its behavior without altering its reactions by one's presence. Bubo was never fully wild, but he was free to come and go as he pleased, and he often ignored me.

Bubo was sometimes a clown, sometimes a terrorist, but first and foremost he was always a very interesting animal. In this book I hope to present him as I saw him from the prespective not only of someone who became very fond of him, but also through the perceptions of a naturalist who studied him.

I GREATLY appreciate the guiding comments and editorial help by Alice Calaprice and Judith May of Princeton University Press. Harry Foster, Mark Konishi, Åke Norberg, Paul Sherman, and Ursula Wartowski also made many helpful suggestions on a draft of the manuscript. Robert W. Nero and Katherine McKeever shared valuable unpublished information on owl molting patterns. I thank the Vermont Institute of Natural Science for taking care of Bubo for one winter. A number of people (named in the book) suffered indignities caused by Bubo. I take full responsibility and ask their forgiveness. Finally, I am indebted to Ruth Goodridge and Erika Geiger for their patience and perseverance in typing the manuscript.

Camp Kaflunk, Maine
November 1986

ONE MAN'S OWL

One touch of Nature makes the whole world kin.

—*Shakespeare,* Troilus and Cressida

Wild Owls

By mid-March in Vermont the snow from the winter storms is already compacted by the first thaws, and the cold nights cap the midday thaws with a solid crust that you can walk on without breaking through to your waist. The maple sap is starting to run on warm days, and your own blood quickens.

Spring is just around the corner, and the birds act as if they know. The hairy and downy woodpeckers now drum on dry branches and loose flakes of maple bark, and you hear purple finches singing from the spruces. This year (unlike most others) the reedy voices of the pine siskins can be heard everywhere on the ridge where the hemlocks grow, and you hear again the chickadees' two-note, plaintive song. Down in the bog, the first red-winged blackbirds have just returned, and they yodel from the tops of dry cattails. Flocks of rusty blackbirds fly over in long skeins, heading north.

From my vantage point at the edge of the woods overlooking Shelburne Bog, I feel a slight breeze and hear a moaning through the forest behind me. It is getting dark. There are eery creaking and scraping noises. Inside the pine forest it is becoming black, pitch black. The songbirds are silent. There is only the sound of the wind above a distant honking of Canada geese from the now starry skies. And then I hear a booming, hollow "who-who-whoo-who—." The deep, resonating hooting can send a chill down any spine, as indeed it has done to peoples of many cultures. But I know what the sound is, and it makes me feel warm inside.

I RETURNED to the same woods in early April. Hepaticas were now spreading their pale blue blossoms over the damp, brown leaves. Crows were starting to build their nests in the hemlocks and pines, and the first spring peepers were calling out, now that the ice had begun to melt. I briefly saw a dark silhouette gliding silently over the pines at dusk; then it vanished like a ghost. But in an instant I knew I had seen a large owl, possibly the great horned owl I had heard the month before. The woods then took on a new meaning: somewhere within this wilderness a pair of the great birds might have its nest. I envisioned the downy chicks for whom this hunter of the night might be foraging.

Of the owls, the great horned owl (*Bubo virginianus*) is the predator supreme. It is one of the so-called "eagle" owls, and among the predatory birds of North America it is exceeded in weight only by eagles. (The European eagle owl, *Bubo bubo*, is larger.) Sometimes it hunts by day, and it is the only owl that has been officially recognized by many state and by federal regulations for falconry.

Ernest Thompson Seton, the nineteenth-century Canadian naturalist, writes about this owl (1890): "My ample opportunities of fully observing these interesting birds in captivity as well as in a state of freedom, and indeed all that I have seen of them—their untamable ferocity, their magnificent bearing, their strictly carnivorous tastes—would make one rank these winged tigers among the most pronounced and savage of the birds of prey." I saw neither savageness nor ferocity in the dark silhouette of the owl gliding by at dusk. It is a predator, to be sure. But is it any more "savage" than a swallow that gulps its mosquito prey alive, or a raccoon that munches on live frogs?

The "winged tiger" vanished behind the pines, but not from my mind. I returned on April 17, hoping to see it again. The crows were now incubating their eggs, and the eastern blue birds were back searching for nesting holes near fields and farms. I crossed a field full of matted dead grass honeycombed with the runnels of meadow voles and entered a stand of mixed evergreens. Here, in the dark and shady glade, raccoons had left their claw marks over the years on the furrowed bark of the thick hemlock trees. The lower branches on these hemlocks were dead, having long since been shaded out by the dense crowns. Owls would surely like to perch here in the gloom to patrol the shady ground below. Indeed, I found some owl pellets—the compacted undigestible fur, feathers, bones, and insect exoskeletons of past meals that owls, hawks, crows, and many other birds regurgitate—confirming that owls had been or were still here.

A great horned owl liquefies a swallowed mouse in five minutes (Grimm and Whitehouse, 1963). Ten minutes later its muscular stomach has wrung the liquids out and passed them into the small intestine while retaining the undissolved content (Reed and Reed, 1928).

Roosting tree in hemlock woods

Pellet formation and pellet regurgitation follow (Kostuch and Duke, 1975). The process from ingestion of prey to egestion of pellet involves seven sequential steps (Rhoades and Duke, 1977) requiring eight to ten hours to complete. On the average only one pellet is produced per day (Marti, 1973).

There are some who cannot resist picking up an owl pellet with its bits of bone, fur, teeth, and feathers and taking it on as a puzzle. Each pellet is a mystery, and behind it is the drama of a predator lurking in the night. Not many decades ago the analysis of owl pellets occupied a small army of biological detectives who busily collected evidence to determine whether or not certain species were "vermin" or "useful" animals. In those days farmers kept their chickens in the open, and the working definition of "vermin" was one who had eaten chicken, partridge, or rabbit. "Useful" birds ate rodents and insects. Owls are more susceptible to incrimination than hawks because they destroy less of the evidence; they have a less acidic stomach than hawks and thus dissolve fewer of the bones they ingest (Duke, Evanson, and Jegers, 1976).

The art of owl-pellet analysis has advanced and flourished. A species of rodent new to science was recently discovered through its remains in an owl pellet (Schlitter, 1973). Fossil owl pellets have also been studied to reconstruct extinct faunas (Dodsen and Wexler, 1979).

Owl pellets — one with remains of a crow's foot. The other with parts of rodent's skull.

David W. Steadman and colleagues (1984) at the Smithsonian Institution in Washington, D.C., relying on fossilized remains of owl pellets deposited in a cave on Antigua, determined that 33 percent of the small birds and mammals there became extinct shortly after human occupation some 3,700 years ago.

Large land animals are much more susceptible to human activities than are small birds and mammals, and major extinctions had previously been noted at the end of the Pleistocene for continental North America, but it was generally believed that the major cause of the extinctions was climatic change. Thanks to owls, we can now separate some of the effects of climate from those of people. New evidence suggests that humans could have been a major cause of the large Pleistocene die-off. It is ironic that owl pellets were first used by people as evidence to incriminate owls, and now pellets from extinct owls are turning up evidence that incriminates people.

As I pulled the tightly compacted fur and feathers apart, I did not need to be an astute detective to recognize the crushed skullbones of rabbits, squirrels, meadow voles, and muskrats, and the feathers and bills of birds. Especially obvious were the feet, bills, and feathers of crows; crows constituted a significant portion of this owl's (or these owls') diet.

Somewhere ahead, I heard loud cawing. The cawing became more raucous as more and more voices joined the chorus. Stalking as quietly as I could toward the commotion, I saw the crows perching here and there in the trees, flicking their wings and tails and making enough noise to mask the sound of my footsteps on the moist pine and hemlock needles. In their preoccupation, they either did not see me or ignored me. Through thick evergreen branches I caught a glimpse of the object of their undivided attention: a large owl with ear tufts—a great horned owl. For an instant our eyes met, and then the big bird turned its head and launched itself over the forest, trailing a rapidly growing pack of excited crows like a ragged plume of black smoke. I was excited. Seeing an owl was better than seeing owl pellets. But finding its nest would be better still.

Dozens of nests of all sorts abounded here, and any number of them could have been an owl's nest. The one that ultimately riveted my attention was high up in an ancient white pine about one meter in diameter and with no solid limbs for at least 15 meters. I could not climb this tree to see inside the nest, but as no pinpoint of light was visible through it, I concluded it was most likely an occupied nest. It was a nest built by crows, but I rejected the idea that it was a currently used crow's nest because of the whitewash that splotched some of the branches—usually a sign of young raptors. Because hawks do not have young this early in the season, this nest was most likely used by an owl. Furthermore, the owl was probably a great horned owl because the other local woodland owls (the saw-whet, screech, and barred owls) nest in hollow trees (long-eared or great grey owls do not nest locally). Great horned owls also nest in hollow trees when they can find a suitably large cavity. But these owls are so large that suitable cavities are scarce, and they have had to make do with other nesting sites. Like their relatives who still nest in tree holes, great horned owls have never learned to built their own nests, and so they now occupy the already-made nests of hawks, crows, and ravens.

Crow parents do not allow the feces of their young to accumulate beneath the nest. After being fed, a baby crow or jay (after the fashion of all other passerine or "song" birds) almost stands on its head while defecating, and the adults catch the feces in order to eat them or dump them elsewhere so that no fecal matter remains in or under the nest.

The young defecate right after being fed, while the parents are still at the nest, ensuring that the droppings can be cleaned up immediately. Nest hygiene is a matter of life and death because the fecal wastes mark the location of the defenseless young for hungry nest predators. To a powerful predator like a great horned owl, however, there is little danger should a squirrel or pine marten decide to sniff out the source of the nest droppings; the owl parents would intercept any such plans.

The nest's surroundings bore other evidence of the presence of owls. There were pigeon feathers, a crow's wing, a flicker's wing, a rabbit's tibia bone with the foot still attached, assorted feathers of songbirds, and pellets like the ones I had seen before, containing crushed and splintered bones of various kinds. By all accounts, I was under the nest of a powerful bird. Then, as if to consolidate my wonderings about the pellets, the feces, and the identity of the nest, a great horned owl sailed out of the tree, flew silently into a neighboring pine, and glared at me.

I longed to look into the nest, but I had reasons not to try. Great horned owls defend their nests fiercely. Arthur Cleveland Bent, in his *Life Histories of North American Birds of Prey* (1961), wrote:

> The behavior of Great Horned Owls in the vicinity of their nests varies greatly with different individuals, though it is generally hostile especially when there are young in the nest. . . . Once I was savagely attacked, while I was climbing to a nest in which the eggs were hatching. I had hardly climbed four feet on the big pine tree, when the great brown bird glided past me and alighted in a pine beyond. There she sat, glaring at me, swaying from side to side, her wings partly spread, her plumage ruffed out, looking as big as a bushel basket, her ears erect, and snapping her bill furiously, a perfect picture of savage rage. As I continued upward her mate soon joined her. . . . Once, when I was not looking, I felt the swoop of wings, and a terrific blow on my shoulder, almost knocking me out of the tree, and I could feel the sharp claws strike through my clothes. . . . As I neared the nest; I felt a stunning blow behind my ear, which nearly dazed me . . . her sharp talons had struck into my scalp, making two ugly wounds, from which the blood flowed freely. This was the limit; I did not care to be scalped, or knocked senseless to the ground, so down I came, leaving the owls the masters of the situation.

Bent's experience was not unique. Donald J. Nicholson (1926) received even rougher treatment when he climbed to within 2 meters of a great horned owl nest containing eggs. He wrote: "Swiftly the old bird came straight as an arrow from behind and drove her sharp claws into my side, causing a deep dull pain and unnerving me, and no sooner had she done this than the other attacked from the front and sank his talons deep in my right arm causing blood to flow freely, and a third attack and my shirt sleeve was torn to shreds . . . tearing three long, deep gashes, four inches long; also one claw went through the sinew of my arm, which about paralyzed the entire arm."

Charles R. Keyes (1911), after receiving a blow from a great horned owl, wrote: "It came absolutely unexpected and was so violent as to leave the left side of my head quite numb. . . . The slash which began on the left cheek and ran across the left ear was rather ugly but not dangerous."

Not wishing to be slashed from cheek to jowl just then—especially while precariously perching in the top of a pine tree—but still eager to look into the nest, I compromised and climbed a neighboring tree and used it for a lookout.

I was charged with adrenalin, and I needed all of it to make it up the pine, which had disturbingly brittle limbs on the lower third of the trunk. Meanwhile, the crow alarm had already been sounded, and crow reinforcements gathered around, quickly coming from several directions. The owl, whom I presumed to be a female because it stayed close to the nest, paid them no visible attention. She remained menacingly close to me, continuing her wide-eyed glare while snapping her bill and hooting, "Whoo-who-who—," and occasionally making hoarse, vehement "wac-wac" calls. When she flew to another tree, she was trailed by at least twenty crows who kept diving at her as if to threaten her with harm. As soon as she landed, however, the crows kept a respectful distance, which happened to be no closer than 2 meters.

The necessity to guard the nest, especially from crows, is great. Owls sometimes do not fare well against crows especially if humans are also present. Frank and John Craighead (1969) describe an episode of crow harassment at a great horned owl nest that did not end well for the owls:

After weighing a freshly hatched owlet and a pipping egg on a cold day in mid-March, one of the authors descended the nesting-tree

and concealed himself several hundred yards away to await the female owl's return. Within 25 minutes she flew back to the nest with
a dozen crows chasing her. Satisfied that the young owlets would
not freeze to death, the observer arose to leave. The owl spotted the
movement and immediately departed, with more than half the
crows in pursuit. The remaining crows did not detect him, so he
continued to watch the nest. Four crows flew at once to the edge of
the nest and then to a tree near by. Nervous and wary, they jumped
from limb to limb, looking at the nest and cawing loudly. One
Crow, bolder than the others, flew three times to the edge of the
nest, only to leave immediately on landing. The three other crows
cawed loudly, as though giving encouragement, but made no move
to fly to the nest. It appeared that the crows had a definite project in
mind and were summoning courage to carry it out. Finally the bold
one flew to the nest, bent down and picked at one of the owlets, flew
off, returned again, and quickly and deftly threw an owlet out of the
nest with a flip of his bill. As the Crow returned to repeat the performance, the observer interfered. The owlet dropped 40 feet but
was unharmed. There was a bruise on the tip of the wing where the
Crow had seized it. Afraid that the performance would be repeated
if he replaced the owlet, the observer carried it to the laboratory and
fed it egg yolk and sparrow liver before returning it to the nest late
in the evening of the same day.

Several days later he returned and took measurements of both
owlets, but on his departure the crows again gave chase to the adult
owl. He suspected that the crows might repeat the earlier performance, so he returned to the nest the next morning and found both
owlets dead at the base of the tree, each with a bruise on the wing
tip as evidence of how it came to its fate. The crows made no attempt to eat the owlets.

There is little doubt that crows have a legitimate reason to dislike
owls. But why do they go to the trouble to mob an *adult* great horned
owl whom they cannot hurt? The question had only been of academic
interest to me before, but actually seeing the crows in action, and the
owl's seeming lack of response, now sparked my curiosity. Small
birds that cannot even make a pretense of harming owls also vehemently mob them. What do they get out of it? I later found many theories in the literature (see Appendix 2), but I wanted to see for myself.

Such a project outside my own area of expertise might be fun, I thought, because I had no ideas or expectations of what the results were "supposed" to be. That meant I had to be open to everything, whether or not it seemed relevant at the time. Not until three years later did I integrate some of the published ideas with my own observations to produce a perspective in my mind (Appendix 2).

From my vantage point I could now see that the old crow's nest that the owls had expropriated was a shambles. It was already a year old and had been battered by winter storms. The owls had not improved it. They had completely battered down its rim, and three fuzzy owlets (two is the usual number) were fully exposed. Two were probably all of 30 centimeters tall, the third smaller. Although all three were still enveloped in fluff, their bills and talons already looked capable of doing damage, and I doubted these chicks were still defenseless enough to be in danger from crows.

There was no fresh food in the nest, only the remnants of past meals—crow wings and bare bones—all without any meat remaining on them. Compared to food provided by owls at other nests, the hunting of these owlets' parents did not appear to have been very good. (Bent [1961] reports that an occupied great horned owl nest, when examined, was stocked with a mouse, a young muskrat, two eels, four catfish, a woodcock, four ruffed grouse, one rabbit, and eleven rats, the whole provender totaling eighteen pounds.)

When food is available only in limited amounts, as is commonly the case in nature, the practice of some owls is that the largest owlet gets fed first. If the last owlet to hatch from a clutch of eggs is able to feed regardless of the competition from its stronger nest mates, it survives; otherwise it is neglected and dies or is eaten. This practice ensures that at least some of the offspring survive when food is scarce. The prospects in this nest did not look good for the runt. Siblicide occurs regularly among some owls and other predatory birds (Andersson, 1982). James Alder (as quoted by Ingram, 1959), who had constructed a blind near a nest of short-eared owls, wrote:

> There were now only five young in the nest, two having previously disappeared. Three of these were quite sturdy; the other two were smaller. After about an hour of watching, the large birds became very restless and presently one of them reached out and pulled at the head of the small one. It then picked it up and attempted to

swallow it head first. This horrified me but I realized in a flash that here was the explanation of the seemingly inexplicable reduction in the size of the family and also in one of the previous year, which had fallen from six chicks to three. I therefore restrained an urge to rush out and interfere in what was happening. The larger owlet presently dropped the chick but after a few minutes again reached out and picked it up by the head. This time it succeeded in swallowing it, and after much writhing and gulping, only its feet showed.

I do not know if young great horned owls are subject to a similar fate. Owls are not all alike. In all owls, however, the young feed on what they find in the nest, and when they are of different age (and size) the opportunity for siblicide exists if the mother needs to leave the nest to hunt. Eggs in a clutch are laid on successive days or every other day, and the female starts to incubate as soon as she begins laying so that the young are hatched on different days. If push comes to shove during food competition at the nest, the first bird to hatch is soon larger than the rest and has the edge. In most other birds not subjected to violent vagaries of food supply, incubation is delayed until the last egg is laid. In this way the parents prevent one offspring from getting a head start and dominating the others.

There are other means of matching the number of mouths to feed to the available food supply. For example, when food is scarce at the beginning of the breeding cycle, great horned owls do not breed at all, or like snowy owls they may adjust their clutch size to fit the food supply (Griffee, 1958; Adamcik, Todd, and Keith, 1978). Snowy owls in Barrow, Alaska, do not breed when lemmings are scarce, but will lay up to two dozen eggs per clutch in peak lemming years, and as few as two in moderate to poor lemming years.

As I continued to observe the nest, the parent owl continued to snap her bill, to call in a hoarse gurgle, and occasionally to hoot while staring at me with her huge yellow eyes. She made no attempt to attack. Her mate did not show himself. Apparently the male had left as soon as I had come into sight, and he stayed away while the female alone kept a continuous watch. (The crows kept me informed of his whereabouts.) These responses are similar to those reported for nesting ravens in eastern Washington (Knight, 1984). Ravens, like owls, are no match for humans with guns, and in range land, where humans are few, ravens dive at people thirty-nine times more frequently and come thirty times closer on each dive than at nests in the more populated farmland areas where they have experienced firearms.

EVENTS showed again how much of what really makes a difference in individual lives is often a matter of random chance, at least to one of the three young owls.

Two days later the skies turned dark as storm clouds came from the north. The wind stopped, and in the hushed silence sticky, wet snow began to fall. It clung to the pine needles, the twigs, and the branches. And it continued to fall. Slowly the accumulating frosting depressed the branches. Lower and lower they sagged, until the stillness was shattered as brittle pine limbs, loaded with soggy snow, came crashing down. I anxiously returned to the nest. It had not gone unscathed. The nest tree was damaged, and tangles of branches lay beneath it in piles of snow. Miraculously, two young owls were perched on some fallen branches. Where was the third?

I had little hope of finding the youngster alive in this snow, but I thought it was worth a try. I pulled out one branch. Nothing. Another. Still nothing. But . . . was that a foot? No doubt about it—an owl's foot was sticking out of the snow. I brushed the snow aside.

The owl attached to that foot was too weak to stand. It was a soggy, sorry-looking bundle of misery, if ever there was one. The excavated bird lay on its side, head hunched down between its shoulders like a turtle in its shell, and it moved its stubby wings weakly in slow motion, like a chilled slug. Only the eyes showed signs of life. Its partially open bill was testimony to its fright, but no sounds emerged. I picked up the limp, pathetic creature and wrapped it in my jacket. There was nothing to do except take it home. Perhaps I could save it. But to take a young owl—or for that matter, a robin, blackbird, or crow—from the wild is against state and federal laws. Perhaps I could study the development of the owl's hunting behavior and determine how or whether its hunting techniques had a bearing on the often vehement mobbing behavior of crows and other birds. Such scientific inquiry would justify my saving this owlet in the eyes of the law.

Owlhood is not likely to be served by ministering to an owl. Helping an owl affects one owl, and that is all. One can help more owls by buying an acre of forest and keeping it wild than by preserving all of those who run afoul with fate, or with civilization. But you do not see the former, or at least you cannot touch them and point your finger and say, "Yes, I helped *this* owl." The unseen, statistical owls are all too easily neglected.

I believe that the accelerating erosion of our natural world can ultimately be traced to our inability to see statistical owls. We are mesmerized only by the real ones. In this book I necessarily focus on one of the latter, but I hope it will not be at the expense of the former.

The academic justification of my taking this very real owl was, of course, only part of my reason for wanting it. In fact, it was only an excuse, and my deeper, fundamental motive was less lofty. Quite simply, I had an attack of what Edward O. Wilson calls "biophilia," a deeply rooted affliction of many of us in the human race who have received a strong dose of nature in their formative years.

We are social animals. We like to feel a part of something of beauty and power that transcends our insignificance. It can be a religion, a political party, a ball club. Why not also Nature? I feel a strong identity with the world of living things. I was born into it; we all were. But we may not feel the ties unless we gain intimacy by seeing, feeling, smelling, touching, and studying the natural world. Trying to live in harmony with the dictates of nature is probably as inspirational as living in harmony with the Koran or the Bible. Perhaps it is also a timely undertaking.

What can transcend the beauty of wild woodland flowers on a sunny spring morning, of geese migrating to the tundra marshes, of an owl gliding through the wilderness at sunset? Because of my knowledge of evolution I know that I am kin to all. I am the accumulated information that has been passed down since the dawn of life, and my body contains the substance that I have inherited from millions of lives, from every species that has ever existed. Knowledge unites.

One's feeling of wonder stems from the perception of incredible complexity in nature, and from knowing that all living things have been shaped over incomprehensible spans of time by relatively simple as well as by dimly perceived forces. The curious child who wishes to pet a wild creature probably wants to do so for many of the same reasons as the scientist who examines it. Though the scientist may appear to have more admirable motives, the child deserves the same freedom to touch wild things and thereby learn to comprehend its place in the natural order of the world.

The touching of nature is, to me, more than a satisfying of my curiosity. It is the source of my wonder. Any one species is a link to my life and to all of life that has ever been. To contemplate the meanings of life is akin to a religious meditation. Indeed, to have a false view of nature is, to me, a sacrilege, because it can breed needless pain. To be a part of nature, through touching and understanding, is to strive to understand the principles of how nature operates. And to know those principles is to be able to work for the good of all. For example, to recognize the natural functioning of germs is to make it possible to cure some diseases; to know our natural psychological tendencies is to be able to create situations that might curb them; to realize that populations naturally increase to their utmost limits is to be able to take countermeasures and to forestall environmental degradation and animal extinctions. Almost all of our problems have a biological basis of one sort or another, and if living in harmony with the other animals in our planet's ecology were a central issue of faith, then paradise (within certain constraints) would be more than a fantasy.

But for the time being, paradise may be out of reach. And so, it seemed, might be the owl, when I considered the bureaucratic impediments that thwarted my legal adoption. But there are reasons for these impediments. Taking an animal from the wild is something one does not do casually. It requires much time and commitment to live with another creature, and one must be prepared to provide not only for its physical needs but also for its psychological requirements. With a fellow human, we can take intelligence and understanding for granted; we can verbalize our feelings or make symbolic gestures to define our relationships. With animals, however, we can never assume that they have the sense or ability to tell us what they need or want. We must study them closely for signs of their needs, and then we must make ourselves available to minister to those needs. The animal lives one day at a time and depends on its human "master"; and so to have one is a constant commitment.

I chose to make such a commitment with this owl, and I decided to clear the appropriate paperwork through the state and federal bureaucracies to make my acquisition legal.

Because I was unable to determine its gender, I refer to the owl as "he" or "him" throughout this book out of arbitrary convenience.

An Owl in the House

REVIVED by the warmth of a wood-burning stove, the owlet stands up in his grass-filled cardboard box, glares at me, opens his bill, and hisses. I hold a piece of meat in front of him but he ignores it. But when I shove it into his partially open bill, he holds still for a few seconds and then swallows ravenously. Each succeeding piece of meat goes down faster than the first, and all signs of meekness begin to leave him.

Well fed and alert, he stands up tall and starts to clack his bill defiantly by snapping his lower mandibles up against the upper. His fluffy down has dried, and he looks twice as big as before. As he spreads out his wings and raises his back feathers, he makes himself look even larger. He faces me belligerently and continues to clack his bill while rocking from side to side, missing no trick to make himself look as large and menacing as possible. All the while, his eyes do not leave me. I recall that the scientific literature about the great horned owl describes it as being fierce, defiant, and untamable, even when young. This one seems all set to fill the bill.

His large eyes are in the front of a flat face, not at the sides as with other birds. His eyelashes are prominent. Where one would expect a nose on a human face, he has a curved bill with a nostril on each side. Most of the bill is hidden by stiff, hairlike feathers, the rictal bristles, which look like they could serve, like a cat's whiskers, as tactile probes in the dark. His quiet dignity, combined with occasional outbursts of ferocity, does not fit one's usual image of a bird.

"Bubo," as I decided to call him, is no longer an infant. Standing flat-footed, he is 31 centimeters tall. But his feathers are still short. His tail, for instance, is a mere stub with only ½ centimeter of feathers at the end of the quills. His wing feathers extend only about 2 centimeters beyond the quills. The talons, in contrast, are already well developed; each is 2 centimeters long, and sharp. As if poking through wool socks, they extend ridiculously far beyond the ends of his toes, so delicately clothed in cream-colored down. Hanging over the legs and extending to both sides, the long, fluffy belly feathers take on the appearance of bloomers. If clothes make the gentleman, then feathers make the owl. He does not look like a dapper gentleman to me, but he could surely pass for a strange caricature of an old man.

His garb is highly functional: as long as it is dry, it helps to reduce the energy needed to keep his body warm. Young owlets are covered with a one-centimeter-long white natal down almost from the time they hatch. At about three weeks of age this neonatal down is replaced by a longer, buff-colored, soft and fluffy secondary down. It grows from the same feather quills as the natal down, which continues to adhere to the tips of the secondary down.

A family of owls nesting during the coldest part of the winter in New England faces a conflicting set of problems if energy supplies are limited—and they usually are. One potential evolutionary strategy to beat the odds in the energy crisis is for the young owlets to be fed by two hunting parents, and to expend some of the energy derived from so much food by shivering to keep warm. Alternatively, another strategy dictates that the young use as much energy as possible for growth rather than for shivering; only one adult hunts, and the other stays at the nest to incubate the chicks. Great horned owls opt for the latter strategy. The female stays at the nest, doing double duty of protecting the young from predators and keeping them warm. The male is the sole provider for the whole family until the young are quite large.

Bubo is only three to four weeks old, but he will probably not gain much more weight. Well-fed young owlets increase their weight by as much as 9 percent each day for the first twenty days after hatching, which means that the demand for food at the nest greatly increases with each succeeding day. The young achieve nearly full weight at only three weeks of age, still six to seven weeks before they are ready to fly. But they do not begin to generate much heat for thermoregulation until two weeks after their major growth spurt (Turner and McClanahan, 1981). Thereafter the mother broods them less and begins to hunt to help feed their rapidly growing appetites. The young can keep warm on their own long before they can fly also because they are tucked into a thick coat of downy plumage. Feathers used for flight are grown after those used for warmth.

Bubo is now too old to be brooded, but I still have to carry out various other chores as substitute parent. And he is not cooperating. He is at eye level in his box on the table in front of me, and whenever I budge he immediately hunches down, spreads his wings, and puffs himself out, rocking menacingly from side to side. He also hisses and snaps his bill in a rapid series of clacks that sound like dry sticks whacked together. His gleaming eyes stay fully open and follow my every movement.

I cut up a mouse and put a small piece into his hooked bill, which he still holds partly open in fright. He clamps down, hesitates a second, and then swallows and closes his bill. As soon as I gently withdraw my hand, he is on full alert again and assumes a defensive posture with blazing eyes and snapping bill. Again I place a small piece of meat in his bill and talk softly to him. The wild look in his eyes subsides and his bill closes again—until I make a small movement. We repeat this sequence at frequent intervals.

APRIL 24

After three days of being constantly beside me Bubo calms down sooner, not only in response to food but also to my voice. After I talk to him in gentle, soothing tones for about two minutes, he begins to smooth his feathers, close his bill, and retract his outstretched wings. I keep talking, and he gradually closes his eyes. However, when I turn the page of a book or lift a pencil he still awakens immediately and acts as wild as before. I talk to him again, and he again closes his eyes. I do not know what he hears when I tell him to relax, but talking gently does seem to affect him. My voice must calm him, because when I stop talking to him, he immediately becomes wide awake and alert.

I want him to ignore me, or to do what he feels like doing without having to take my presence into account. I am therefore elated when, after four days, for the first time, I see him rise up to his full height and leisurely stretch out one wing. He would not have allowed himself such a trivial gesture unless he had, at least temporarily, considered me only a trivial presence.

For the time being, however, my presence is still a big factor in his behavior. He is even alarmed when I turn a page of my book: the motion causes him to hunch his head down and alternately hiss, clack his bill, and vibrate his throat (gular fluttering). Nevertheless, at times he does ignore me enough to focus his attention on a page of the book I am holding. For now the paper has begun to become more threatening than I am.

Gular fluttering is a rapid movement of the skin in the throat that results in a bird's taking shallow breaths without hyperventilating the lungs. Ordinarily it is a mechanism by which birds, including the great horned owl (Bartholomew, Lasiewski, and Crawford, 1968), move large volumes of air over the moist membranes in the throat so that excess heat is lost through evaporation (birds do not sweat). Is the gular flutter, when used in fright, an anticipatory response in preparation for a fight or for flight, to eliminate heat quickly? We have a similar heat-dissipating response when we are nervous: we sweat.

Bubo with a deer mouse

APRIL 26

After five days of being with me constantly, Bubo is now downright tolerant of me. By moving my hand very slowly I can even reach around behind him and scratch the back of his head. He responds by shaking his head, as if he were trying to get rid of a bothersome bug. I scratch some more, and he ignores my fingers probing his soft and fuzzy head. Strangely, when I gently touch the feathers of his belly, he starts to groom his wings, drawing the feathers through his big hooked bill. Then, for good measure, he precariously balances on one foot, closes his eyes, and scratches the feathers around his eyes with one of the great talons of the other foot. The preening completed, he fluffs himself out and shakes himself violently, throwing a shower of white flakes—the dried remnants of feather sheaths that are split and shed as the feathers grow—in all directions. Finally, he rises tall on his fuzzy legs, stretches one wing and his neck, and exhales with a gentle sigh.

Later with eyes still fully closed, he raises his head as if in alarm and begins gular fluttering. Has he just dreamed about a frightening experience, such as when he was swept to the ground and buried in an avalanche of snow? His eyes remain closed, so he cannot be frightened by something he is seeing. When he is awake, he immediately looks intently at anything potentially threatening. But no, he is not awake now. Bubo is dreaming!

I am attributing human characteristics to Bubo when I say he is dreaming—an act we call "anthropomorphizing." Such attribution causes dreadful outrage in many who have some knowledge of biology. But the issue is not as clear as it seemed to be a few years ago. "Anthropomorphizing" can mean many things. If scientists use anthropomorphisms, they are doing so merely as colorful allegorical devices to illustrate a scientific point. Most will agree that a bee stores food "for the winter" even though it has no conscious knowledge of the changing seasons. Therefore we can safely say that bees store honey for the winter because we are referring to the *evolutionary* reason for food storage (that is, those bees that stored food lived and multiplied, and those that did not, died). On the other hand, when we say that a dog buries a bone "to hide it," it is indeed proper to wonder whether or not the dog knows what it is doing. Until recently, even though a pet owner may have insisted that his or her animal has humanlike qualities, many biologists considered nonhuman animals, whether bees or dogs, as reflex automatons; to endow animals with the ability to feel or dream was to anthropomorphize. New developments in biology, however, show that we have much in common with some of the "higher" animals—maybe even with an owl.

Bubo, who stood up during his "dream," is now lying down again. With eyes still closed, he makes a very distinctive great horned owl hoot, "whoo-who-who-who-whoooo—." It is a soft and reedy, whispered sound, coming from deep down in his throat, and I have to be very close to hear it at all. I am amazed by it. Most birds are not capable of singing their species-specific song until they reach sexual maturity, and even then many have to learn it from their parents while young. This owl has just given a good rendition of its species' song long before even leaving its nest.

April 28

Like most felines that abound by the dozens in the barns of many New England farms, Bunny, my wife Margaret's cat, *catches* more than his fair share of songbirds, yet he *eats* mostly dried cat chow and tuna. I put a dead hermit thrush that Bunny has caught into Bubo's nest. Bubo has eaten birds before. The remains under his former nest attested to that, but he completely ignores the thrush. Would his parents still tear the prey apart for him? Perhaps he still does not recognize food unless it is handed to him properly. I will soon find out.

Before I probe my pupil, I want to transfer him out of his nest and onto the arm of the sofa beside me. It is not easy. Bubo vehemently dislikes anyone laying hands on him. But there is a way. I raise my gloved hand under his breast, and he climbs on, snaps his bill a few times, and clumsily shifts his weight from one foot to the other. His policy seems to be that if you cannot get out of the way, then get onto it. Although this is the gentlest way of moving Bubo, I now have the problem of getting my hand back for other uses. It takes a bit of twisting to get him off, and Bubo maneuvers like a lumberjack riding a log on a river drive, but not successfully. Eventually he is perched beside me on the arm of the sofa, with a newspaper strategically placed under him. The questioning begins: Does he notice the whole hermit thrush at his feet? Apparently not. Does he recognize the bird as food when it

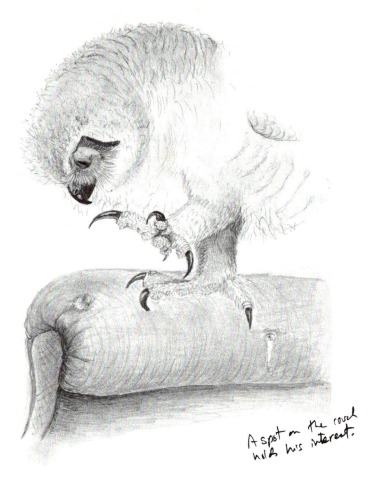

A spot on the couch holds his interest.

is cut up and handed to him in small pieces? The answer is a resound-ing "yes."

Bubo shows contentment after feeding by grinding his lower man-dible along the upper while drawing his tongue in and out. This makes a soft, crackling, popping sound quite unlike the defensive bill snap-ping. He also fluffs out and shakes himself, and seeing his comfort-signs I feel happy, too.

After a good meal Bubo smacks his bill in contentment for two to three minutes. And then, apparently having forgotten his meal just past and not yet thinking of his next meal ahead, he shifts his atten-tion to his environment. Turning his head around slowly—right, left, and back again—he methodically inspects the walls, the floor, the ceiling, and then he peers out through the window. But not for long. Next, the woodstove seems to mesmerize him, and inexplicably he becomes animated. He rapidly bobs his head up and down and to the side and back, a good 15 centimeters each way. The bobbing helps him perceive depth; closer objects will appear to move more than distant ones. With lightning swiftness his great pupils dilate and contract. After a frenzied ten-minute inspection of the room while he totally ignores me next to him, he makes a decision: he leaps 45 centimeters off the sofa onto a chair. He barely makes it with his feet, but his heavy body lags, so he dangles from a talon or two and weakly flaps his ineffective wings. I help him onto the chair while his talons draw a few drops of my blood. Then he hops clumsily onto the floor.

Bubo walks on his oversized feet by hunching down and leaning forward, taking long, slow, and deliberate steps while his wings are pulled in tightly. A ballerina he is not.

To babysit an owlet is a chore, and you must do it diligently, unless you like your house whitewashed. Margaret has hang-ups about that, and so I retrieve him to the chair each time he hops to the floor.

MAY 4

Bubo is already showing his hunting instincts. He has watched a beetle scurrying on the floor, although he did not attack it. Now he makes three pounces in a row, each time "capturing" a leaf or a piece of straw, which he grasps in one of his huge taloned feet. He lifts the foot to his face and painstakingly transfers the object he has captured to his bill. He displays amazing persistence at his game, being totally undistracted by my laughing and talking.

How much might his natural behavior change by his association with me? The debate of nature versus nurture is an old one. But it is off the mark. There is no nature *versus* nurture. One is not possible without the other. Life is like a fire that consumes resources and changes them to something else. Genetic capacity is like a spark or flame. But the size of the final flame is not related to the initial spark; it is determined by the environment—by the substrate availability and by constraints such as temperature, wind, moisture. Bubo's hunting skills are, like a spark, already contained within him. Given the availability of prey, he will learn to hunt it, like he will learn to fly, and there will be little I can do to either hinder or facilitate his capacities. All he needs is opportunity.

Bubo is now beginning to exercise his wings in earnest (until now he has only stretched them frequently). Perched on the arm of the chair, he repeats about ten rapid beats every 15 minutes or so. He already rises slightly into the air, but his talons still keep him securely anchored. Undoubtedly he will need to practice more than foot-and-wing coordination to be a good hunter. He may have to learn about the prey itself. But for that he will have to be in the wild, perhaps in the woods near my camp in Maine.

A wild predatory bird is not likely to act naturally toward prey when it is being watched by a human observer. I hope Bubo will eventually be tame enough to trust me so I can see his natural behavior in the wild. There will surely be a trade-off: after having been tamed, he may no longer act like a "real" wild owl in all things, but I should be able to learn something, nevertheless; if he were totally wild, then I would likely not be able to see any of his natural behavior.

Bubo lying down, and watching a grasshopper.

Owl at Kaflunk

BUBO's cardboard nest is closed, and the box sits on the back seat of my jeep without much jostling. Packed beside and around it are blankets, sleeping bags, groceries, books, papers, instruments, tools, dishes, utensils, and all sorts of other paraphernalia needed not only for my survival in the woods for four months, but also to allow me to conduct field research there. Bubo and I are going to Maine, to Camp Kaflunk, for the summer. And for some time the two of us will share the camp alone. Margaret may not be able to join us for another week, or maybe two.

Camp Kaflunk lies nestled in a small clearing next to ledges atop a hill called Adams Hill. The clearing is surrounded by white pines, red spruces, and grey and white birches. The nearby mixed forest grades into red and sugar maples and beech and descends to swamps covered with thick sphagnum moss. The swamps are bordered with white cedars and balsam firs. The description of the Maine woods by Henry David Thoreau back in 1846 still fits best: "It is a country full of evergreen trees, of mossy silver birches and watery maples, the ground dotted with insipid, small red berries, and strewn with damp and moss-grown rocks, . . . the forest resounding at rare intervals with the note of the chicadee, the blue-jay, and the woodpecker . . . ; and at night, with the hooting of owls and howling of wolves; in summer, swarming with myriads of black flies and mosquitoes, more formidable than wolves to the white man" (Thoreau 1972). The only difference between then and now is that instead of the howling of wolves one now hears the howling of wolf-coyote hybrids.

Kaflunk is a deer-hunting camp that was built by Bucky and Edna Buchanan more than forty years ago, but the rough-sawn 2 × 8 beams and unplaned pine boards are as solid now as they were then. The cabin measures 3.6 × 6 meters in the interior, quite luxurious dimensions for a hunting shack in Maine. A wooden ladder leads up to a small loft. Under the loft I have placed a pile of sawed firewood and a

Camp Kaflunk.

bed; rough shelves for books are on the wall. At the opposite end of the cabin is a cast-iron sink. My desk is in between, close to a small cast-iron stove and a low, wooden table.

From the western window I can glimpse a distant lake through the trees. Mt. Tumbledown and a series of other steep, ledge-topped hills rise beyond it. No houses or camps are visible in any direction.

The narrow trail to the camp winds along an old stone wall, outlining the edge of an abandoned apple orchard now being taken over by ash and maple trees. Near the apple orchard are remnants of an old field—a reminder of the old farm of Asa and Elmira Adams, for whom this hill was named.

The Adamses and their descendants, the Yorks, were born on this hill and spent their lives here. They raised sheep and cattle, but mostly they cultivated apples. They grew "Ben Davis" apples, a variety no longer grown commercially. In the fall these apples were stored in the barn with hay on top of them. Now the only reminder of the barn is a foundation of stones overgrown with feral roses and raspberries in a clearing a short way below Camp Kaflunk. But some of the apple trees survive. Bears now climb the remaining gnarled, hollow giants. They pluck green apples directly from the branches and eat them on the spot. Deer, ruffed grouse, and porcupines also eat the apples after they fall and are tinged with frost.

Only a cellar hole full of fieldstones remains of the old homesite, plus a couple of barn foundations and two old wells. We have repaired one of the stone-lined wells, and we draw our water there.

Edna, a descendant of the Yorks, had visited her grandparents here when she was very young. She told me some of her memories about the place: "I thought it was the most beautiful place in the world. The house had clapboards, but no paint. My grandparents were very poor people. The house had two fireplaces, one with big Dutch ovens. Grandmother decorated the walk with wagon hoops, and she always planted flowers. There were cosmos, bachelor's buttons, and asters along the walks, and there were two barns. The privy was near one of them, about 100 feet from the house. In those days the land was all cleared, and we had views in all directions."

Edna, now a kindly old lady who is everyone's ideal of a grandmother, also told me, with a gleam in her eye, about the big old crabapple tree by the edge of the field near the well, and the view clear down to the lake. Edna's husband, Bucky, claimed "It's the best place around. There isn't any like it, anywhere." I agreed. And when the land was for sale again (after having been sold once already to a lady in Boston), at one hundred dollars an acre, with Camp Kaflunk thrown in for free, I did not resist the bargain, even though (or perhaps because) the land no longer had a road access.

I have kept the field near the old homesite cleared of the returning forest, and here we plan to build a log cabin out of the spruce and fir trees from the surrounding woods. To me the cabin will not only be a structure; it will also fulfill a dream.

Beyond our field the trail continues up through a stand of white pines, and finally terminates on the ridge top where Kaflunk sits among spruces and birches on moss-covered ledges. Bucky and Edna told me that they built Kaflunk from the remnants of a dilapidated cow barn. A sign with the name "Kaflunk" was nailed over the door by friends after the camp-warming party, before the Buchanans had time to think of a name of their own. And the name (origin unknown) and the sign have endured until now, although the peeling paint makes the letters almost undecipherable.

A country road used by horses and buggies once passed just below the cabin, before industry lured people to the towns and the farmers left this rocky soil to go West. More enduring, however, are the massive stone walls. Although these walls have stayed, most of the fields and sheep pastures they enclosed have been reclaimed by forest, and the sheep have been replaced by moose, deer, and bears. Within the last twenty years, coyote-wolf hybrids (known locally simply as "coyotes") have appeared as well, filling an ecological niche left by the timber wolves that were exterminated soon after the land was settled. Ravens have also come, close on the heels of the coyotes, to feed at their kills. Many wild animals share portions of the trail with us, and their tracks are a visible reminder of their presence.

I discovered these woods twenty-nine years ago when I was an avid young deer hunter. On frosty mornings I would see the tracks of deer beneath the gnarled apple trees along the stone wall, and I would follow them for days on end in the snow to the top of the black ridges and back into the swamps. I also remember seeing the scratch marks of bears on the beech trees, smelling the autumn leaves and the first falling snow, and finally experiencing the deep winter silences. The more I saw on these wilderness jaunts, the more I wanted to come back the next season. Not much has changed. I still see the woods as Thoreau did: ". . . a specimen of what God saw fit to make this world."

MAY 8

Bubo and I arrive at the foot of the hill near Kaflunk in the early afternoon, and I now have to make nearly a dozen trips up and down the steep half-mile trail to provision the camp. During the first trip I transport Bubo. He rides perched on my arm, his head swiveling this way and that. However, we do not get very far before he hops, or rather tumbles, to the ground and starts to run. I chase after him and catch him, again and again. I learn that the higher I hold him up, the less frequently he jumps to the ground, and the more quickly we get to Kaflunk. After having the run of the cabin for a few minutes, he stares long and hard at the top of the woodpile. The woodpile it is. He climbs it, then relaxes there and goes to sleep while I continue making trips to carry up the supplies.

The snow has melted only recently here on the hill. The spring season is not so far along as in Vermont. The insect-pollinated varieties of spring flowers have not yet begun to bloom on the forest floor, whereas in Vermont the trout lilies, Dutchman's breeches, bloodroot, spring beauties, and white trilliums are already in full bloom. Only the wind-pollinated plants have bloomed or are blooming here: the poplars, witch hazels, elms, and red maples. Queen bumblebees are already flying about, and at a large sugar maple along the trail I hear loud humming. The queens of the bumblebee species *Bombus terricola* are feeding at the yet unopened buds, most likely on sugary secretions.

The warblers are returning. Ovenbirds, Nashville warblers, black and white warblers, black-throated green warblers, parula, magnolia, and Canada warblers, and yellow-rumped warblers are already in the surrounding woods, singing and establishing or reclaiming territories. Some are still migrating through the area. A flock of over fifty yellow-rumped warblers comes foraging through the birches, generally proceeding in a northerly direction.

MAY 9

Yesterday evening Bubo retired to the top of the woodpile and went to sleep before dark, and he did not move from his spot all night. In the early morning, as I build the fire and make my breakfast, Bubo's gaze from the woodpile follows my every move. His head turns slowly as he watches me. No excitement, just awareness, maybe curiosity. He is still perched in exactly the same place on the woodpile when I come back to the cabin in the evening to cook supper, after spending the day in the woods. He looks calm and drowsy.

About half an hour later, at 6:30 P.M., however, he becomes quite another owl. Suddenly his head begins to swivel in all directions, and after a few hops he comes down to the floor of the cabin. He also gives a little whisper of a hoot: "Whooo who-who-whooo-whoooo." His head now turns in quick jerky motions, first in one direction and then in another. These head turns are so rapid that the motion itself is hardly visible, only the end result of it.

Having explored the floor, he hops onto the bed, on which there are blankets, pillows, two jackets, and my running shoes. First he attacks the shoes, and then in turn the red jacket, the blanket, and the dark blue jacket.

He is busy at his task for 20 minutes, and judging from his bill-smacking and rapid, animated movements, he seems to be enjoying himself greatly. I call him again and again and wave my arms. Does he pay even the slightest attention to me? No! Occasionally he condescends to stop and look up; but almost immediately thereafter he continues to do whatever he had been doing, such as pursuing a flying moth. In the meantime I am left waving my arms and shouting myself hoarse.

Bubo is a born hunter, and he wastes no time in beginning to practice his craft. As Goethe said, "Whatever you do, or dream you can do, begin to do it." Bubo follows Goethe's advice, and he is not easily distracted, nor is he particular: almost anything is fair game.

I pick up a black and white pen and roll it across the floor. This object does not resemble anything that he has eaten, so he should not confuse it with food; he still eats only what is provided by my fingers. Will he chase the pen as a form of play? It stops rolling, and Bubo stares at it for several seconds. Then, with neck outstretched, he scurries toward it and pounces onto it. He picks it up with his right foot and tries to bite it. He nibbles on it, but it is rather slippery and falls from his bill; it rolls on the slanted cabin floor until it disappears out of sight under some shelves. Out of sight, out of mind? Not with Bubo. After a pause and another look around, he runs quickly to the shelves, trying to pry underneath them with his bill. Sorry. There is no space for this owl's big head down there. I conclude that he chases as a form of play; at this point in his development the chase does not yet have a connection with food.

At about 8:30 P.M. Bubo executes one final attack on the jackets and shoes. Without having damaged them, he jumps up onto his woodpile, pulls his head down into his shoulders, bill-smacks contentedly, and closes his eyes.

MAY 10

Bubo was quiet all night, and this morning he is in exactly the same place where he went to sleep last night. Later in the day he is still there, but he is now lying down like a sleepy dog with his feet stretched out in front of him. His head is resting on his taloned feet, and his eyes are partially open, watching me. Some bird.

At about 6 P.M., as yesterday, Bubo stands up on his woodpile and looks around attentively, beats his wings about a dozen times, whispers a hoot, and finally lifts his bill straight up while he stretches and yawns. At 6:15 P.M. he hops one meter lower to a pine limb I have provided as a step; he exercises his wings once more, and then, five minutes later, he hops onto the floor. Like yesterday, he seems to be feeling well and is excited; he bill-smacks and leaves tiny white punctuation marks on the floor behind him that seem to be forced out of his body by his excitement. He pounces on a knot in the floor, and then onto one of the little white specks he has made. He again hops onto the bed that also serves as my closet and bureau, and here he gives the jackets and the shoes another ineffectual but impressive inspection. Life with him is settling into a predictable and comfortable pattern.

His clumsy play-pouncing before he is even able to fly shows that he is genetically programmed with some hunting responses. His great excitement and drive to explore everything new are probably also adaptations related to hunting, but they suggest an open program appropriate for learning. At this time, however, my speculations are pointless. I hope that Bubo will provide me with some insights to his hunting techniques in the coming weeks.

May 11

As usual, Bubo slept soundly all night. If ontogeny repeats phylogeny, then his ancestors were probably day-active. I adjust my waking-up time to his—6 a.m. sharp—when he hops off the woodpile onto the floor and play-pounces, with spread wings, on a few knots of the pine floor planking. After this ritual he hops back onto the woodpile to retire, presumably for the rest of the day. I go into the woods.

After returning to the cabin in the late morning to escape the freezing temperatures and dreary drizzle, I am cheered by Bubo's companionship. He perches on my arm while I walk around and make myself a cup of coffee; when I sit down to drink it and to read, he amuses himself by nibbling at the buttons on my shirt. But he does not nibble only on buttons. Whenever my hand comes near his bill, he nibbles my fingers, too, and we play finger-bill games for about half an hour.

Does he confuse my fingers with edible meat? To find out I feed him the mice that I trap nightly in the cabin, and the birds and squirrels that I pick up on my daily run on the highway below while training for a big race in the fall. Soon he is so satiated that he ignores the meat that is directly in front of his bill, and I offer him my bare finger, which he again nibbles gently. He is indeed distinguishing between what the fingers *hold* and what they *are*.

An hour later, some of Bubo's appetite is back, and I am ready to explore the same question with a different experiment. This time I hold a piece of birch bark in front of him. Sure enough, he snatches the bark with his bill—and gulps it down. It doesn't go far; he immediately regurgitates it. When he finds birch bark anywhere else, he either ignores or plays with it, but he never swallows it. But when I offer him anything at all, he seems to assume it is something to eat. Bubo, you trust me too much, but I am honored and touched that you have swallowed the birch bark. It proves your trust.

It is raining and sleeting hard, so I sit by the fire and read all morning. I remove one birch log after another from under Bubo and place them in the fire. He has a difficult time settling down, being disturbed so often by having his perches taken out from under him. Nevertheless, he finally lies prostrate on top of a thick log, but his head droops down over the side, and he looks uncomfortable. I decide that he needs a proper bed.

A shallow cardboard box filled with dry leaves and pine needles and placed on the logs does nicely for a nest. Once inside, Bubo shifts his weight on his furry feet, and grabs a dry, wrinkled leaf in the steel-blue talons of his right foot. The four talons lock inward in a viselike grip on the fragile, crackling leaf. He lifts the foot, and quizzically looks at the leaf—and then he nibbles at it with his bill. After a quick shake of the head, he drops the leaf and lies down again and closes his eyes.

At 4 P.M. Bubo is still lying in his new bed, and he watches me intently for several minutes at a time. Then he stands up, stretches both wings by raising them above his head, and regurgitates a huge pellet containing the undigested bones, fur, and feathers of his recent meals. Strangely, he holds the pellet in his bill—and then to my amazement he gulps it back down. [I would never again see him repeat this behavior.] Apparently well satisfied, he now does a series of rapid wing strokes, shakes himself, and nonchalantly preens his wing feathers.

The black sky clears later in the day, and I venture outside the cabin. When I return at 7:15 P.M. Bubo is already hard at play on the cabin floor. He is pouncing from one knot to the next, trying to grab and bite each one, without much success. He then finds a new toy—a 5-centimeter-long stick of pine. It clatters and rolls when it is dropped, and Bubo picks it up, drops it, and picks it up again and again for fifteen minutes. This game seems to get him excited, because he is again expressing tiny white droppings, and they keep getting progressively

Skull of a young barn owl,
showing eye sockets, nares,
and ear openings.

smaller as the effort to produce them presumably gets greater. He suddenly drops the stick and, as if losing interest, starts to walk away.
Then he wheels around and pounces on it with renewed vigor.

After Bubo tires of playing with the stick, I take him onto my arm
so that I can scratch his head, and I observe the details of his expres-
sions. When he is at rest and at peace, the feathers above his bill rise
and make his face appear fluffed out. His eyes are oval, as he partially
lowers the top lids and slightly raises the lower lids. Occasionally one
eye remains open while the other is closed; but when he is at full attention both eyes are wide open, and the feathers of his head are pulled
back, making his eyes appear round and large. When he is alarmed, his
head feathers are pulled back and close-set, but when he is bluffing
they are fluffed out.

May 12

Before Bubo makes a leap, he turns his head as on a swivel and bobs it
up and down like a yo-yo. It all has to do with his large eyes. Indeed,
they are larger than a human's—so large that the margins of the eye

are outside the skull, where they are encased in a protective sheathing of bone. Because there is little room left in the skull for the muscles that move the eyes, he has to move his whole head to scan the area for a potential perch and to gauge the distance to it.

At one time people thought that an owl cannot move its eyes at all, and that they merely stare foreward like headlamps. Nevertheless, at least in the great horned owl, small horizontal and vertical eye movements have been demonstrated in the laboratory (Steinbach and Money, 1973; Steinman, Angus, and Money, 1974). These eye movements are so small, however, that they are probably used only for fine-tuning. Bubo's constant head-turning, therefore, is a consequence of the limitation of his eye movements, which is in turn the result of an adaptation for increasing eye size that is related to vision in dim illumination.

MAY 13

Bubo is obviously greatly excited as I take him for a walk, perched on my arm, just outside the cabin. But it is not the ride on my arm that excites him. In no time at all he hops onto the ground and runs into the woods. He pays no attention to me when I try to call him back, and I must follow him through the underbrush.

It may not be wise to let him loose. He is a long way from being a hunter, and it is hard to predict where he might wander. At the same time, he is now so active in the cabin that, given his size, he is becoming the proverbial bull in the china shop. There could be other complications. The daily bread of adult wild great horned owls consists of rabbits, but the species has, on occasion, also been known to dispatch house cats. Margaret and her orange cat Bunny will soon be joining us. Margaret named him "Bunny" because he reminded her of a rabbit. Will Bubo agree?

I have not dared to discuss the possible cat-owl interaction with Margaret, because she has already expressed some reservations about living in the woods with me and an owl. But that discussion need not yet be broached, because Bubo has yet to dispatch even a mouse. So for now it is unlikely that Bunny is in danger; on the contrary, he is a formidable hunter, and I am afraid for Bubo.

After weighing all of the imponderables, I decide that, for the time being, Bubo should be caged. And so I build an aviary abutting the southern end of the cabin. There is a window between the cage and Kaflunk, so Bubo can be let into and out of the cabin. The aviary contains thick branches, and under one end, under the roof, I have built a large artificial nest of sticks lined with grass and dry ferns. What more could an owl want? I cannot think of anything else to make it a more comfortable home. But I am forgetting Murphy's Law, which says that if everything seems all right, then you obviously do not know what the heck is going on.

May 14

When I put Bubo into the enclosure he hops from branch to branch, and almost immediately clambers onto the nest. He relaxes there, and then sleeps all day. So far so good. At night, though, it becomes another story. He becomes alert at dusk, as usual, but he does not play at all and spends most of the time gazing out into the woods. He spends the rest of the night on a perch, and whenever I look in from my bed to check on him, he is still wide awake.

Having Bubo caged is no joy at all. As he stares out (and presumably he listens, too) into the spruce and birch woods, with all of the wonders in them that he has never explored, he reminds me of a starving prisoner shackled by the ankles and just a few centimeters away from a nourishing meal. It is very likely that I am projecting my own feelings, but I do not think this kind of anthropomorphism will hinder scientific progress or prevent serious inquiry into owls. An owl is a part of the woods. It has evolved in them, and its responses are linked to its natural environment, where presumably it would instinctively want to be.

When I look out my cabin window over the clearing and the forest and over to the ridges I feel good because I can walk out the door if I want to. When I see the raven fly by and hear it call, I can think about where its nest might be and perhaps look for it. What I see and hear thus provides functional connections to my environment. And when my environment is well, then I feel well.

A tantalizing study on some hospital patients suggests that we might be measurably affected by what we see. Convalescing hospital patients have no choice but to be confined and restricted in their environment. Roger S. Ulrich (1984) of the University of Delaware asked patients if what they saw, either a brick wall or some trees, affected them physically. Examining the records of patients recovering from gall-bladder removal in a Pennsylvania hospital, he reported that those patients with a view of a natural scene, in this case the trees, spent on the average one day less in the hospital than those who saw only the brick wall. They complained less and took fewer painkillers. Other studies have already shown that people with pets also enjoy significant health benefits, such as lower blood pressure, a positive attitude toward life, and longevity.

Apparently there is something that we get psychologically from our contacts with the natural world that ultimately helps to sustain us physically. Bubo was also probably programmed through evolution to seek, enjoy, and live in those habitats where he could be best sustained. Anxiety is an adaptive response that gets animals, including humans, out of harmful situations, and in Bubo's case it should be no exception. Like humans, however, he will no doubt cease to struggle when he does not see viable alternatives.

Beginning Hunter

MY LIFE in these woods so far has consisted mainly of studying bumblebees as they forage from flowers of various shapes, scents, and colors, offering the bees their nectar, pollen, both nectar and pollen, or nothing. The individual bees from a colony never know what will be available to them as they start their lives as foragers. But they have evolved an open program of survival whereby they learn how to find and manipulate the different kinds of flowers, and this then allows the different individuals of a colony to forage at any time of the summer, wherever the colony might be. The cost of potentially having everything is that initially they have very little. They make mistakes. At first the bees may visit flowers but get no rewards, they may approach bright objects that are not even flowers, and they may handle some flowers clumsily. Through experience, however, they learn to restrict themselves to the best bargains available, and they become skilled flower handlers at the most common and remunerative food sources (Heinrich, 1979).

I wondered if the great horned owl might behave in a similar way. It, too, has the problem of correctly identifying and handling prey from a vast array of potential prey and other objects in its environment. And the kinds of prey that are available at one place or time of year are often not the same as those at another place or time.

MAY 15

The time passes quickly when you always seem to be behind as the season rushes on. Coming home to Bubo every evening makes the time fly even faster. And now I have another reason to hurry home: Margaret and Bunny have also joined us.

The dreaded encounter between Bunny and Bubo last night was a draw. The antagonists squared off and eyed each other. Bubo lowered

Bubo sitting back on his "heels." Darker new feathers are growing into wings.

his head and ballooned out to bushel-basket size by spreading his wings and raising the feathers on his back. Then he blinked and rocked slowly from side to side. His bill-clacking sounded like two boards being rapidly struck together. Bunny meowed, feigned disinterest, and slunk out of the cabin.

In the morning I let Bubo in to keep us company. He eats a shrew and a mouse that Bunny had caught during the night, and he finishes up by swallowing a chicken drumstick. Since he is not very active during the day, we let him return to his favorite perch on the woodpile after his Sunday brunch. Bunny remains asleep on the mattress up in the loft.

When Bubo sits up there on the woodpile, you notice his eyes a lot. He stretches frequently, turns his head, nibbles on his toes, and preens his back and wing feathers—all with his eyes closed. But when not preoccupied with personal hygiene, he quietly looks at you for as long as you care to look at him. He only glances at Margaret, but he watches me closely.

He blinks in the same way we do (and unlike most other birds), by moving his top lid down across the eye. But when he closes his eyes to snooze he moves his bottom lids up. Initially his top lids move down a little but his bottom lids then push his top lids up, so that the eye is eventually entirely closed by the bottom lids. Both of the top and bottom lids are feathered. In addition, he has a pair of milky white inner lids, the nictitating membranes. These can slide over the bright yellow eyeball, being drawn from the upper inside corner diagonally across and down. Much of the expressiveness of his eyes is not so much from the position of the lids as from the elevation of the feathers surrounding the eyes.

Unlike our pupils, the openings of his are not controlled only by light. His pupils dilate and contract in a flash even when his head has not moved and the illumination has not changed. Sometimes the two pupils open and close independently of one another even when his

Bubo bobbing his head up and down to achieve better depth perception.

head and eye positions do not change. There must be another factor besides light that determines the opening of Bubo's pupils.

Before Bubo jumps to a new perch he looks at it attentively, and the pupils of his eyes contract and dilate rapidly. Closing the aperture of the lens in a camera increases the depth of field, and I suspect that Bubo uses this method in addition to head-bobbing to measure distance (for instance, to a perch).

ACCURATE sight and depth perception are undoubtedly of great importance when pouncing on prey. Hearing, however, may be just as important. Indeed, Roger Payne (1962, 1971) at Cornell University has shown that barn owls are able to strike mice in total darkness, orienting themselves only through sound, and great gray owls catch mice by striking at them through snow after they hear them (Nero, 1980). But vision is required to learn to use the auditory localization. Barn owls raised with one ear plugged can learn to make accurate pounces, but they make systematic errors when the earplug is removed (Knudsen and Knudsen, 1985). The visual fine-tuning of the auditory localization can apparently occur only in younger birds. Those under seven months of age can gradually readjust to again make accurate pounces after their earplugs are removed, while older birds, once they have learned one set of rules about localization, maintain these rules and make localization errors indefinitely when they are forced to hunt in total darkness.

MAY 16

Bubo has been eating all the mice that Bunny has been bringing in at night and depositing under our bed, and I supplement his diet with road kills, mostly birds. Today I bring in a treat—a red squirrel. Will he swallow it whole, head first, as he does the mice? No. He acts as though he has no idea what this animal is. When I hold it out to him he nibbles at it, but that is all. Maybe I should present the squirrel to him in another way.

Before he can assume a free life in the wild, Bubo must be able to pursue and kill his prey. I doubt he is ready to kill yet, but will he pursue? To find out, I tie the dead squirrel on a long string and gently pull it across the floor about 1.5 meters in front of him, so that it is moving away from him. Bubo takes notice. He looks, and lifts his head a little higher and looks some more. Now his head bobs up and down as he tries to get a better distance perspective, and then very slowly he takes a tiny step forward. Just one. Perhaps he does not dare to advance any more than that. I now very gently pull the squirrel the other way—toward him. Bubo elevates his wings in alarm, starts clapping his bill menacingly as he lowers his head, and stares wide-eyed at the thing coming toward him. Then he bolts away, running with rapid steps while bent over like a hunchback. He looks back once over his shoulder, clacking his bill during the hasty retreat. He almost trips over himself, but he catches his balance and escapes.

May 17

Cats are supposed to catch and remove mice from the house, but Bunny brings them in, sometimes in a very lively condition. He brings us not only mice, but other wildlife as well, a favorite being the tiny smoky shrews, *Sorex fumeus*. We usually have two to three of these every night under our bed, but they are always dead. (Undoubtedly the live ones escape through the cracks of the floor.)

I tie one of the many available shrews onto a string and drag it across the floor. Bubo, perched on my wrist, comes to attention. He pounces, and lands on it. Then he brings one foot up to his bill to bite into his prize. But it is the wrong foot; the shrew is under the other foot.

Bubo has a long way to go, but at least he is going. We resume the game: I pull the shrew, making it "escape"; he looks at it suspiciously, runs after it, and pounces on it again; he nibbles at it, but then he drops it. His prey-catching behavior may be fairly innate, but it needs a little touching up. It is still not clear, however, if he associates the moving things he "catches" with food.

For days now Bubo has been looking up at the beams in the cabin. He cannot fly up there yet, but when I hold him up he jumps onto them without hesitation, and then he immediately begins gular fluttering as if he were nervous. Still, he seems to like to be there, and he reminds me of a kid on a scary roller-coaster ride at the fair.

May 18

Bubo does not yet have the power to gain altitude in flight, but he can proceed horizontally with apparent ease, at least for 3 to 6 meters. Still, if he goes anywhere at all, it is generally on foot.

Although he has attained a weight of 1.4 kilograms—nearly his full expected weight—his twelve tail feathers are now still only 9.3 centimeters out of their sheaths. They are still less than half their final length.

Whenever we open the window to his cage he comes into the cabin. He then hops onto the top of the woodpile and lies down on an old shirt. With eyes half-closed, fully closed, or sometimes fully open, he nibbles on the shirt or on his toes, while putting his weight on his "heels" so that his toes are raised. Occasionally he stands up and yanks vigorously at the shirt. Or he stands on only one leg, lifting the other leg and nibbling at his clenched talons. Now a fly comes by. He stops his manual exercises, opens his eyes wide, and watches it while it circles him at a dizzying speed. Bubo moves his head in lightning-quick jerks to keep up with the fly's movement, but even an owl can turn its head only so far, and this one is no match for the frenetic fly. Instead, Bubo decides to watch a bird on the birch tree outside. Anything that moves is of great interest to him.

May 19

After we get up in the morning, he appears at the window, and I open it so he can come in to join us. His stare and his preflight stoop tell me that he intends to fly onto my shoulder. I do not yet have my shirt on. All eight of his 2.8-centimeter talons are needle-sharp, and I take evasive action just in time. He flies across the room and lands on my desk instead, causing the papers to fly like leaves in a windstorm. Now he grasps the back of a chair, hangs on tightly, and vigorously exercises his wings, scattering the papers even more.

I make a fire in the stove, and after the stove is slightly warm Bubo jumps onto it. He dances up and down, beating his wings and snapping his bill loudly. [He never jumped onto it again.]

In the evening he is as active as ever, maybe more so, because now he has discovered a wonderful new mode of locomotion—flight. Papers, dishes—nothing is safe anymore. He has become the feared owl in the china shop, though it is doubtful Kaflunk could be considered a china shop.

He likes it when I run my fingers through his head feathers. That slows him down, but I tire of it after a half-hour, apparently long before he does. Afterward he is ready to resume his mischief, unless I redirect his attention.

I have saved a rabbit's foot from one of Bunny's kills a few days ago.

It is quite dry and hard now, but it might do for a toy. As I pull it along the floor on a string, Bubo watches alertly. He runs after it and pounces with his feet extended forward and his head back, in the classical pounce of owls in the wild (Norberg, 1970). There is only one problem—he lands about 8 centimeters short of the mark. No matter. He looks down and grabs the rabbit foot in his bill, and then he swallows it, trailing a string from the side of his bill like a long, skinny rat's tail. His first catch! And he has connected chasing for fun with food. I let him keep the foot by cutting the protruding string. His hunting fever aroused, he now looks all around and tries to bite into a few of the knots in the wood, which he had ignored for several days. Apparently his search image for prey is still crude, and perhaps he is still primarily motivated by play.

I am running out of entertainments and it is getting dark. Time again to light the kerosene lamp and put Bubo back outside into the aviary. He runs back and forth on the ground along the wire in his cage, his head bobbing wildly as he looks out. Then he runs foot over foot up the side and across the wire screen roof, aided by his beating wings. And then he does it again, and again. Next he flies with full force against the screen, making a distinctive and indelible depression, like a brand, in the soft keratin on the base of his bill where he hit a wire. He is constantly looking around and jerking his head. I want to let him go *free*. But I fear he will immediately run or fly into the woods, and I know he has neither learned to come to my call, nor is he able to fend for himself.

May 20

Bubo's back feathers are now well developed, but they are fringed at their ends with tiny grayish-white tassels of down. These tassels—the remains of his baby plumage—are now wearing off, leaving dark cream-colored feathers underneath. His breast and head feathers are still the off-white nestling down of three weeks ago. Apparently there is no great need to shed them quickly; the priority now lies in the wings. His primaries are almost full-length, and he is already growing entirely new chocolate-brown wing coverts.

His legs are enclosed in a layer of cream-colored fluffy feathers that extend all the way down and over the tops of his toes. The undersides of his toes have huge bumps studded with a dense covering of hard pegs that should act like tire treads in helping him to secure his grip.

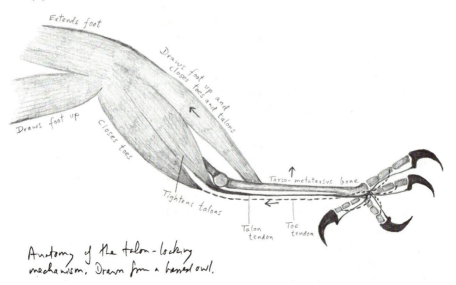

Extends foot

Draws foot up and
Closes toes and talons

Draws foot up

Closes toes

Tightens talons

Tarso-metatarsus bone

Talon tendon

Toe tendon

Anatomy of the talon-locking
mechanism. Drawn from a barred owl.

His talons are curved inward when closed, so that any prey that is grasped and that attempts to pull away will impale itself even further by its own efforts to escape.

I WONDERED how the locking mechanism of his toes works, and my subsequent dissection of a road-killed barred owl allowed me to examine it in detail. Each leg has a tendon that runs down to the foot, where it branches out and attaches to the four toes. The momentum of the impact onto prey or perch causes the legs to be drawn up to the chest, and as the tendons attached at the base of the toes are pulled, the toes curl inward into a fist. To augment this four-pronged mouse-trap, there is a second tendon, even thicker than the first, whose ends attach to the base of the talons. As the legs are drawn up, this tendon is also pulled, and it causes the talons to lock inward; but the automatic locking action of the talons can be augmented by contraction of a powerful muscle in the "drumstick" that pulls this tendon still further. The talons lock inward, and they remain locked as long as the legs are bent. Thus the recommended procedure for getting an owl (or a hawk) off your arm is to first straighten out the bird's legs.

The use of the owl's feet as a lethal weapon is documented by graphic descriptions in literature. Norman Wilkinson (1913) describes an encounter between an owl and a skunk:

The opened talons, as during a pounce, and closed. Drawn from a barred owl.

Mouse traps

One morning, late in the autumn, I was driving [in a horse buggy?] through the woods, when I heard a disturbance in the dry leaves at a little distance from the road. . . . As I drew near, I saw clearly the cause of the disturbance. A few feet in front of me was a large horned owl in a sort of sitting posture. His back and head were against a log. His feet were thrust forward, and firmly grasping a full-grown skunk. One foot had hold of the skunk's neck and the other clutched it tightly by the middle of the back. The animal seemed to be nearly dead, but still had enough strength to leap occasionally in the air, in its endeavors to shake off its captor. During the struggle, the owl's eyes would fairly blaze, and he would snap his bill with a noise like clapping of your hands. Neither the bird nor its victim paid the slightest attention to me though I stood quite close. How long the owl had secured the death grip I do not know, but there was no doubt about his having it. The skunk could no more free itself from the owl's claws than it could have done from the jaws of a steel trap. Its struggle grew less and less frequent and at the end of about fifteen minutes they ceased altogether

Bubo's feet look formidable. They *are* formidable. But for the time being he probably does not know his own strength. It will be some time before his feet will be used as lethal weapons. For now he uses his talons to delicately scratch the feathers around his eyes or on the back of his head, or to pick up objects to bring to his bill.

MAY 22

Last night we allowed Bubo to stay inside. He spent the entire night perched quietly on a 2 × 8 beam near the door. This morning he is awake at 5:15 A.M., stretches, looks around, and does his wing exercises.

Each day his wing exercises become more vigorous. This morning I count six exercise bouts in all, each for several seconds and spaced out over half an hour. His training reminds me of the interval work I used to do on the track. The wing-beat frequency during these sprint exercises is now three per second. Not bad for a bird like Bubo with a four-foot wingspread.

Margaret suggests I put Bubo outside into his cage before we have breakfast. I try. I carry him to the window and extend my arm out as he is perched on it (he does not tolerate being held), but he walks right up my arm, back into the cabin. Several such trips are enough for me. I do not care to press the point. All right, Bubo, you win. He takes his victory nonchalantly and looks up to the rafters, so I hold him up and he hops onto them. He does one more set of wing exercises, and then stands on his left leg, tucking the right as a clenched fist deep into his breast feathers. He closes his eyes. To avoid trouble later I put newspapers on the floor beneath. Our breakfast is peaceful. Bubo spends the whole day perched in the same place.

I still do not dare let him loose outside into the surrounding forest because I am afraid that he will disappear and starve. I may be totally wrong; perhaps he would explore only around the cabin and then stay close by. But I do not yet know enough about him to give him that chance.

May 23

There is something unique about Bubo's wing feathers. They are covered with a soft downy vellum, and the small vanes jutting off the leading flight-feather shafts are curled so that the feather edge is not sharp as it is in most birds. This feather construction probably reduces the noise of flight, so that the prey will have difficulty in detecting the approaching owl, and also so that the owl hears less interfering noise when it locates its prey by sound during flight.

At least to my ears, Bubo flies silently. In contrast, the male woodcock we recently heard at night has a special stiff wing feather that makes a specific whistling sound in courtship flights. Of course, most birds have some whistle to their wings; for example, one can hear ducks flying from a long distance. But Bubo must fly silently to approach and catch alert prey. He is a model of functional design, from the fine details of his feathers down to the tips of his toes.

I admire Bubo's thickly feathered feet and legs as he perches on my

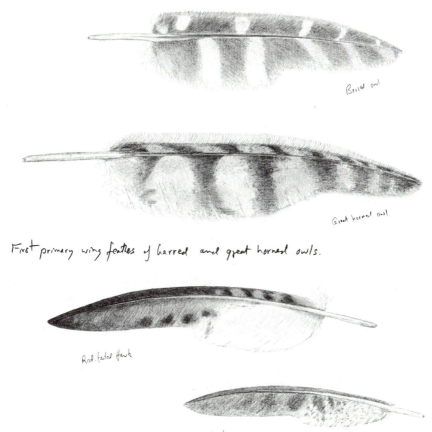

First primary wing feathers of barred and great horned owls.

First primary wing feathers of red-tailed and goshawks.

Wing of
Barred Owl
(Length = 36cm)

Shape of a
falcon's wing

Outline of an owl's wing, compared
with that of a falcon.

hand. Songbirds, which catch their food with their bills and use their
feet strictly for perching, do not have feathered feet. Why does Bubo
have such thick stockings? From the point of view of warmth, it
makes sense because he hunts in the northern winter by sitting stolidly in the cold for days. He kills prey by the strength of the grip in his
feet, and those digits must be quick and strong at all times. The cold
would make them weak and uncoordinated, like my fingers get when
I am out in the cold for a few minutes without mittens. His thick
down "skirt," into which he tucks at least one foot while he perches,
also undoubtedly helps to keep his feet warm.

STRANGELY, although the owl's feet are important weapons, its powerful bill, which is so effective in tearing up prey and crushing bones,
is apparently seldom used in the excitement of battle. S. A. Grimes

(1936) describes an encounter between a great horned owl and a large, 115-centimeter-long black snake:

> [The owl was] lying on its side with its wings outspread, trying its best to get its talons on a black snake that had coiled around the bird's abdomen just back of the breast bone and beneath the wings. The snake had gotten itself around the owl in a double coil. Six or eight inches of its head and neck and perhaps a little more of its tail were free, but the bird appeared to try only to get hold of the strangling coils around its body. It is easy to see that the bird could not possibly get its talons on that part of the snake tightly wound around its abdomen, but why the hooked bill was not brought into play is hard to understand.

I am getting very fond of this bird, and it will be difficult for me to leave him at the end of the summer. But if we take him with us back to Vermont I will have to keep him in a cage. And if I plan to keep him in a cage for the rest of his life, I probably should not let him out now to taste the freedom of the great outdoors.

As we discuss Bubo's fate, he hops onto the bed where the cat is napping and stands up tall and stares at it. The cat stares back, and a staring contest begins. Who will back down first? The cat just lies there, but after ten seconds he displays hints of nervousness by twitching his tail. Bubo continues to stare. After another twenty seconds Bunny suddenly bolts from the bed and runs outside through his "door," the hole in the broken window.

During the last night Bunny brought back a young rabbit, and as usual he only ate the head. I hand the rest of the rabbit to Bubo, who begins chittering excitedly, and with his eyes closed he nibbles all over it. He always closes his eyes when he puts his bill near food or prey. Is this an adaptation that prevents eye injury caused by struggling prey? After he has palpitated his rabbit with his bill, and massaged it with his toes, he drags it into the darkness under the bed. There he begins tearing off chunks to eat. He eats all but the back and hindquarters, which he pulls and pushes into the farthest, darkest corner under the bed where he has never ventured before. Then he comes running out.

Evening. Bubo has slept well all day up on the beams. He stands up, stretches, yawns, and then in one leap hops down to the floor. He lands with a loud thud, and without a pause he runs directly under the bed to the far corner where he had in the morning hidden the hind-quarters of the rabbit.

He seems to know that the confining area under the bed can be a dangerous place to be in. Bunny walks near the bed, and Bubo, although he has finished eating the rabbit by now, becomes defensive, spreading himself out, rocking from side to side, and clacking loudly from under the bed. Later, when he is out of the corner, and safely perched above Bunny, he gives him scarcely a second glance. The distance between himself and the enemy is not so important as the strategic position, which he clearly appreciates. Can he visualize consequences before having experienced them?

Bubo threatening Bunny.

Bubo stretching by
lifting both wings.

Bubo stretching by extending
left leg and left wing down.

Freedom

THE cabin has always been crowded, and now it seems to be getting even more so as Bubo becomes increasingly active. We do not want to keep him cooped up in an aviary. He can fly well and should be able to escape ground predators such as the cat or raccoons. So, we will soon give him his freedom, and whether or not he stays with us will be up to him.

MAY 24

The door is open. Bubo walks to the portal, bobs his head excitedly as he surveys his new environment, and then he hops down into the grass. Not even the young rabbit I hold in front of him distracts him. Instead, he flies from the ground straight up onto the cabin roof, and there he stays perched all day. Occasionally he walks or flies back and forth along the ridgepole. I no longer worry that he will try to make a quick escape. If he goes anywhere, it will probably not be far.

Because Bubo is so visible on the cabin roof, I think that he will be mobbed immediately by all the neighborhood birds. But I am wrong. Pairs of slate-colored juncos, hermit thrushes, and white-throated sparrows have nests with eggs in the blueberry patches near the cabin, and a pair of robins has a nest with eggs in a spruce 60 meters from the cabin. But none of these pairs, nor others who are also incubating or have offspring nearby, pay any attention to him. In the evening a blue jay comes by and screams at Bubo, but only briefly. Meanwhile, a heavy rain begins to drum on the roof, and the downpour continues all night. Bubo becomes soaked in minutes, and he pulls his wings close in to his body.

MAY 25

Bubo is still perched on the ridgepole this morning, but he looks a bit bedraggled. His normally fluffy breast and head feathers are matted together in wavy strands, and after he shakes his head he has a tufty reggae look. We call him Reggae Owl. I try to tempt him back down by waving the rabbit at him several times, but Reggae Owl just looks. Finally, in the evening, he launches himself from the roof, and in a graceful flight lands 3 meters from me on the ground. He approaches and I give him his first bite, and then I back up to encourage him to come to me again. Does he? No! He sees me dangling the meat, but he stays rooted to the ground and bites and nibbles at the leaves and sticks at his feet. Maybe, according to him, where he has received one morsel of food is where there are others, even though his eyes must tell him differently. I give up, finally, and bring him the rabbit. His appetite is good.

After his meal he immediately resumes to perch on the roof, and then he flies to the large white birch in front of the cabin window. I guess he does not yet care to go exploring. Instead, he sits facing the evening sun, with his eyes closed.

A blue jay comes by again today but leaves after giving only a few warning cries. Later, a blue jay perches 3 meters above Bubo in the birch. This jay gives no warning cry at all, but Bubo clacks his bill at him. A flock of fifteen to twenty late migrant warblers stops to forage in the same birch he is perched on, but none of these pay him any visible attention. I am puzzled.

May 28

It poured without any letup last night, all day yesterday, and the night before then. Throughout this deluge Bubo has done nothing but sit stolidly in the birch, only occasionally shaking his head to maintain the reggae look.

Now, at 7 A.M., we awaken without hearing the pounding of rain on the roof. Birds are active again. Another flock of migrating warblers, vireos, and flycatchers is foraging in the branches of the birch all around Bubo. The tree has no leaves yet. The birds can't avoid seeing him. As before, as far as I can tell, these birds just ignore him. And I had thought that all birds mob owls vehemently. I have much to learn.

June 1

Over the last two days Bubo has established a routine. It is simple: he sleeps on the roof, hops down to the ground in the morning to be fed, flies back onto the roof, and then flies onto a large limb of "his" birch, facing Kaflunk. Another significant thing in his life is his toy, the remains of an abandoned anthill that is overgrown with moss. He plays with it after being fed, and before flying back onto the roof.

He attacks the moss ant hummock with an energy I have never observed in him before. In his haste he tumbles over onto his side, and sometimes he even rolls on his back. He strikes at it again and again with his talons, driving them in deeply, tearing out tufts of moss with his talons as well as with his bill. He occasionally stops and looks all around, and then repeats his attacks. How would he fare against real prey?

Young predators generally start with small prey, and as they get older and more proficient they move up to larger animals. Most predatory birds, including large owls, choose insects as their first live prey. Small insects generally do not fight back, and the birds gain experience and confidence. Moss hummocks may sharpen the use of legs, wings, talons, and bills in assault tactics as no insect ever could. This hummock does not fight back, and all of the assaults are successful. With Bubo, nothing succeeds like success, and nothing ensures more failure than failure.

ONE of Paul Errington's captive juvenile great horned owls (Errington, 1932) struck and killed a young cottontail rabbit on its very first attempt. Rats were then "taken care of with equal effectiveness and more precision." But a guinea pig that darted around the cage erratically drove the bird up on the wire of the cage in fright. The guinea pig incident ruined the confidence of this owl as a killer. Thereafter it was afraid even of the rats it had previously dispatched with ease. Errington said, "Especially was this true of rats that advanced; rats that retreated stood an infinitely better chance of finding an owl on top of them." Another of his owls that was fully adept at catching rats was also demoralized by a darting guinea pig, and only after five days of fasting did it again attempt to kill rats released into its cage.

JUNE 2

For several days blue jays have been coming to pick up scraps of meat I leave out for them. They ignore Bubo for now. But the robins, who have a nest with four eggs nearby and who totally ignore Bubo, instead

Bubo in sliding flight.

vehemently scold the jays. Jays are notorious egg thieves, unlike owls, which eat only meat, possibly including young and tender birds. Can the robins actually know this? They act as though they do. [A few weeks later, when Bubo and I approached a robins' nest with ready-to-fledge young, the adults mobbed him continuously for a full seventy minutes until we finally left.]

THE blue jays eventually raided the robins' nest. Apparently the robins had indeed correctly appraised their enemy. Was that from previous experience or from an innate recognition of their enemy? The latter seems like an inadequate explanation when the same bird will mob very dissimilar nest enemies. I have seen red-winged blackbirds, for example, chase wildly after nest raiders such as ravens, crows, and broad-winged hawks, and they mob humans as well.

JUNE 6

Bubo has now enlarged his world to include the surrounding trees, nevertheless he continues to stay quite close to the cabin. My fears that he would get lost in the woods were unfounded. If I cannot see him, I call him by name, whereupon he wheels into the clearing. It is a beautiful sight. With a few pumps of his wings, he approaches a heavy limb of the birch from underneath, sets his wings into motion, and then, banking up sharply, he slows his momentum and makes a gentle landing. Often he lands on the rocky ledges where he plays with me, and then stretches out prone like a cat to rest in the sunshine.

Sometimes, when he is playful, he grasps my ear in his talons, and I cannot pull the offending foot away unless I want to risk getting my ears perforated. He plays rough, and so do I, but eventually he tires of it and lies down in my arms. Looking at the clock I see that we have played for one and a half hours. It seemed shorter than that.

In the afternoon a colleague who also does research on insects comes by for a visit. He is our first human visitor on the hill this summer. Bubo sits on the screen door and looks down at him inquisitively.

Finally he comes down to me but he keeps his distance from the stranger. Bubo seems to be very nervous, and at first he does not even eat the mouse I offer him, as he continues to eye Ken. But he does have his limits. He gulps the mouse and then makes a quick getaway back into the woods.

JUNE 7

Bubo does not seem to be anywhere near the cabin this morning. This is very unusual. Can his absence have anything to do with Ken's visit yesterday? As I am about to give up calling, he arrives but is not in a good mood. As usual, I put my hand under his toes to lift him up onto my glove, but today he snaps his bill loudly and vehemently. Strange. He is not threatening me, because these bill-snaps are delivered with his head held high and his eyes clearly averted from mine. In his intimidation displays, his bill-snaps have a more hollow sound, and he lowers his head, spreads his wings, and pointedly glares directly at his adversary.

This morning Bubo is a different creature from the playful owl of yesterday, when he was on my lap and allowed himself to be stroked. Could he indeed be upset about yesterday's visitor? Ken is of approximately my size and build, except he has red hair and I have brown. Bubo can apparently tell people apart, but it is not only by their exterior appearance. His reactions to me are the same regardless of what clothes I wear. Just to test him again I now put a pillow case over my head to hide my face, and I notice no change in his behavior. There are mysteries here, and I do not yet have the answers.

Bunny caught another young rabbit last night, and although I am not at all pleased by his depredations, I have no choice but to tolerate them. The surrounding farms have cats breeding freely in the barns, and these cats forage in the fields and forests. The cats' owners need a hunting license to kill a rabbit, and they would get fined if they killed a warbler. But one can have unlimited numbers of cats who can kill for you by proxy all they want, every day, in hunting season or out, protected species and unprotected species—and no one cares. Should they?

Predators, including domestic cats, concentrate their activity usually on the most common species; that is, they generally prune the excess population of any one species before it threatens to harm the environment for other species. In this way the diversity of species is maintained, and the predator's overall role is beneficial to the populations it preys on (Tietjen, 1985). But the domestic cats replace the natural predators who would otherwise be able to live off the prey. That includes statistical owls, and owls are federally protected birds.

Bubo is pleased to get the rabbit for breakfast. He feeds from it for awhile, and then flies off, trailing it along in his talons. He carries it about 30 meters into the woods, and then drags it with one foot while doing a one-legged hop with the other. Eventually he shoves it with his bill underneath a brush pile.

In the evening he returns to the cabin, and, once again his usual playful self, he attacks the moss hummock. While he is preoccupied I go to the brush pile to check on the rabbit. Until now Bubo ordinarily did not follow me, but he now comes swooping down toward me just as I near the brush pile. Not only does Bubo remember where he had hidden the rabbit, he also seems to check up on it when he sees me in its vicinity. He peers at me a long time. I wonder why.

Later on, when he is again at the cabin, his pupils contract to near pinhead size as he stares fixedly into the sky. I look, too, and to my surprise—it is late in the season—I see a formation of Canada geese flying north at a very high altitude. They appear only as a series of tiny dots, but to me these moving dots are of interest not because of what I see, but because my imagination tells me what they mean. Bubo presumably does not see beyond the image. He does not know about geese, about migration and navigation, about the fluffy goslings these birds will rear, about the physiology of flight at altitude, or the reasons for the V-formation. Yet his eyes look on, and he gazes after them as if they were of a great interest to him until they disappear beyond the distant ridges.

JUNE 8

I give Bubo his first run-over snake chopped into several sections, thinking he might not recognize as edible a long thing that resembled a smooth rope. Or, if he did consider such a thing edible, would he know how to handle it correctly?

I need not be worried. He eats every piece, and I then give him a dead, but intact, garter snake. It feels smooth and oily to the touch when rubbed in either direction. Though the snake is long dead, Bubo bites into it, making crisp little crunching noises as his bill cracks the vertebrae.

The snake was already limp, and now it is even limper, so it can be curled up in a tight place, such as in Bubo's crop. He picks it up by the tail and starts gulping it down. After about five gulps he has swallowed 40 centimeters of snake, with only a little over 15 centimeters more to go. Then he seems to realize he has done it all wrong. He stretches his neck, opens his mouth wide, and out comes the snake. He picks it up again. Now Bubo does it the "proper" way. The proper way to eat a snake is to start at the *head* end and to finish with the tail. I would not have thought it mattered, really. But apparently Bubo has certain rules of etiquette. They were not learned from me.

LATER in the summer I introduced Bubo to live garter snakes by releasing them in front of him in the clearing. He backed off a little from his first encounter when the snake coiled, flattened itself, and made wide lateral body movements. It lunged at him with its pink mouth wide open. Taken aback a bit, Bubo hesitated, but then hopped right onto it. The snake did not take kindly to this maneuver and bit into Bubo's foot, locking on. Bubo then jumped up in alarm and flew off, dragging the snake that conveniently relaxed its bite to release itself near the edge of the woods. There it escaped. The second snake began with the same defensive maneuvers. But after eyeing it for a minute or so, Bubo finally struck it with both feet simultaneously and, almost in the same instant, jerked his head down, grabbed the snake's head in his bill, and crunched it. He swallowed the snake almost immediately—head first, of course. Snake number 3 made a short sprint into the grass, with Bubo running behind it. Apparently aware of the close pursuit, the snake suddenly stopped dead-still and wiggled the tip of its tail. Bubo seemed mesmerized by the twitching tail, which gradually stopped moving. He then looked all around, as if searching for the "real" snake. Not seeing it, he walked to the tip of the tail. The snake now tried another sprint, but this time Bubo pounced, and snake number 3 met the same end as snake number 2.

So far Bubo has caught no live prey. Today marks the beginning of this new endeavor, on something less threatening than a snake—an insect. He is perched on my wrist, idly biting the leather glove, when he abruptly stands tall. I hear it, too, at about the same time he does—a soft rustle near the edge of the cabin. An investigation turns up a luna moth fluttering in the raspberry bushes. Bubo watches it for thirty seconds, hops down from my hand, and in one quick strike of his left foot he has it in his talons. Next he bites into it, making a crackling, crunching sound, and then he swallows the huge pale green papery moth in one gulp. He ends the small meal with his little bill-smacking noises of satisfaction.

JUNE 9

Some of the glacier-scarred granite ledges near the cabin are bare, although most are overlain with a thin layer of soil that supports moss, lichen, and stunted spruce trees. On one of the bare ledges, 9 meters from the cabin, is a bathtub-sized depression in the rock that catches the rainwater. This pool is like a small oasis up here in the woods and is used by many birds for drinking and bathing. In the summer the water is bright green with algae, and it always produces a crop of tadpoles. I do not know how the wood frogs manage to find this small basin atop the ridge, but they invariably do. Dragonflies also lay their eggs in the pool, and their larvae eventually eat many of the tadpoles.

We used this water before we cleaned out and relined the old farm well near the cellar hole of the old homestead in the clearing below. Periodically I scoop out all of the water to remove the decaying leaves and algal growth. In a few days, after being filled following heavy rains, the pool begins to be repopulated—with water striders, water beetles, mosquito larvae, and of course green algae. Now, with well-water available, we leave the pool for the frogs, the insects, and the birds.

This morning, under a bright and sunny sky, Bubo hops down to inspect the pool. He peers into the water, cautiously walks in, dips his head down. He makes little smacking-sucking noises as he brings his head up again and swallows. As far as I know, this is the first time he has ever taken a drink in his life. I am sure he does not need to drink even now, since he gets adequate moisture from the meat he eats. Birds do not excrete urea, like we do, which requires much water and is flushed out in urine. Instead, like insects and lizards, they convert their nitrogenous wastes to uric acid, which can be excreted as a white paste with very little water lost in the process.

Bubo continues to walk deeper into the pool, slowly and with mincing steps, and as he wades deeper he ruffles his feathers and shakes his wings lightly, tentatively. And then he takes the big plunge. After completely dunking his head, he thrashes his slightly outstretched wings so vigorously that the water foams and sprays in all directions. Again and again he dunks his head completely under and beats his wings, churning the pool like a rotary eggbeater. All the while he makes squealing noises unlike any he has ever made before. When he finally walks out of the water he is thoroughly soaked, and water runs in small rivulets off his matted breast feathers onto the ledges. He looks at the water again, walks in, and repeats the whole performance. He does it three times: his first bath is a thorough one. Afterwards, he flies onto the cabin roof and sits in the sun on one leg with his wings drooping. After an hour or two of preening, he is restored to his old fluffy self.

JUNE 10

This morning, while we are having breakfast, I hear Bubo clacking loudly outside for brief, intense bursts lasting a second or two at a time. Strange—the cat is still sleeping on our bed. I rush outside just in time to see a broad-winged hawk swoop over him, coming within centimeters of the top of his head. The hawk then perches in a nearby tree, eyeing Bubo. I am reminded that a broad-winged hawk, possibly this same one, has a nest with four eggs less than one kilometer from the cabin in the fork of a maple. The hawk makes another swoop, again divebombing within a few centimeters of Bubo's head. I step into view, and the hawk disappears.

A little while later Bubo is ready to play, nibbling at Margaret's shoes while she is outside sitting on the moss in the sunshine. Every time she moves he chases after her, pouncing on her feet. She calls to me to bring her the glove we use for him to hop onto so that we can bring him up to eye level to talk to. Margaret gets the glove, but right after that I see her chasing Bubo across the clearing, and *Bubo* is carrying the glove in his talons. She catches up with him, and then I hear a scream. Margaret comes back with the glove, but her fingers are dripping some blood. She does not give up the glove easily, nor does Bubo, as I would expect from a predator who has to catch and hold prey that is often large and defiant.

In the afternoon I have a treat for Bubo—a dead red squirrel. I hold it up for him to see, and he leaves the moss hummock he has been attacking and comes to me on the run. He bends forward when he runs, and waddles a bit from side to side, which is not considered good form by the running cognoscenti, but he makes good time nevertheless. He grasps the squirrel with an unbridled enthusiasm bordering on ferocity, putting two toes of one foot forward and two aft. (Unlike most perching birds who have three toes forward and one back, his fourth toe can be placed either forward or back.) Then, having a solid grip, he proceeds to squeeze the squirrel violently with both feet in synchronized contractions, as if trying to kill it. In the wild an owl would very rarely take dead animals. Perhaps that is why Bubo treats the dead animals as if they were alive.

Bubo has no way of knowing that this squirrel is not going to be "killed" so easily. It is attached to a fishline, which I yank as soon as he has relaxed his grip. The squirrel escapes his grip. He gives it a brief, puzzled look, then runs after it and catches it again. We repeat the game several times. Bubo is definitely no longer afraid of the retreating red squirrel and I feel encouraged by his progress. However, to really put him to the test, I pull the squirrel so that it moves toward him. He raises his head in surprise, takes a few steps backward, and flies away in apparent fright. So—back to smaller prey. I release a running ground beetle (Carabidae) near him. I wager he cannot resist a small fast-moving object. He does not. He is next to it in a bound or two, picks it up in his bill—crunch, crunch—and then he swallows it, noxious defensive secretions and all.

The vigor of Bubo's play with us is now often intense, and it is mostly at our expense. Flying adeptly, he comes out of nowhere to pounce on our feet or to land on a shoulder, where he is apt to start tugging at an ear. If he nibbles with his bill we can always pull free because he never bites down hard. But when he uses his talons, he reflexively holds on more tightly when he gets excited, and when we try to brush him off he gets excited. Because of the wonderful fish-hook locking mechanism of the talons, we don't try to extricate ourselves from his grip. So, outwardly we try to be very calm and cooperative with Bubo to avoid any dangerous entanglements.

The Gourmand

UNTIL very recently, raptors who ate rodents were considered to be "good" and worth protecting, while those that ate other birds were considered "bad" and had a bounty slapped on their heads. Most states protected all owls except the great horned owl, because although most owls eat primarily mice, the great horned owl does not. Its pellets show that it feeds on snakes, fish, frogs, insects, and various small mammals and birds. The mammals on its menu include hares, cottontail rabbits, squirrels, rats and mice, muskrats, gophers, weasels, mink, skunks, woodchucks, opossums, porcupines, domestic cats, shrews, and bats. The birds it eats include grebes, ducks, chickens, turkeys, Canada geese, bitterns, herons, rails, snipe, woodcock, quail, grouse, and doves; harrier, Cooper's, red-tailed, and red-shouldered hawks; other owls, flickers and other woodpeckers, jays, crows, starlings, blackbirds, snow bunting, juncoes and other sparrows, mockingbirds, and robins. Despite this variety of fare, Errington (1932), in his study of the great horned owl in the north-central states, found that the staple diet over most of its range consists of rabbits. Bubo's diet was also broad, as it largely reflected Bunny's hunting success and the vagaries of automobile kills. However, he did have quirks and special likes and dislikes.

JUNE 12

The sluggish stream in the valley one kilometer below Kaflunk is well stocked with clams and minnows. Both seem especially common at our swimming hole. I have cleared away the alders around the pool, and the bank is now covered with bracken fern and grass. The pool itself is at the base of some riffles, and it is still partially paved with rocks. It was once paved with many more and much larger jagged rocks, but I throw a few out onto a crude dam every time we come down for a dunking. Now there are already some cleared patches where clams are moving in. They insert themselves half-way up their shells into the sand and silty mud, and then they leave weaving tracks behind them as they slowly pull themselves along the bottom. Most of the clams are aligned with their partially open valves facing upstream, from where they filter the water that comes from the beaver bogs above. The water slows down in the pool, and silt then settles out. The pool would eventually fill with silt if it were dammed long enough. But I destroy the dam every fall so that silt collected over the summer will be swept out of the pool with the spring torrents. In the summer silt that collects is stirred up only by our walking, and schools of minnows gather to nibble near our toes and on them. If it turns out that Bubo likes clams or minnows, he will have a constant feast.

I crack a clam with a rock and offer Bubo the contents. He takes the slimy mollusk reluctantly, then holds it immobile in his bill, as if wondering what to do next. Eventually he swallows it. So far so good. I offer him a second one. This one he refuses. He learns quickly—too quickly to suit me. Maybe I can fool him. Margaret has just brushed Bunny, and she gives me a fistful of his orange fur. Clams wrapped in cat fur—now that's more like it. More roughage. Bubo gulps several furry clams with gusto. Apparently he is more concerned with texture than with taste.

JUNE 13

I sit down near the pool on the ledge, and after I call Bubo, he walks over and stops. He does not look up at me, but down into the water. Strange—what could he be watching in the water? Then I look into the water, too, and find myself looking into his eyes through the reflection. He sees me in the water. Nothing strange there. I move. He looks up, and now he also sees me on the opposite side of the pool *out* of the water. Instant alarm. There is supposed to be only *one* of me here! He jumps up and hastily retreats into the woods. In a few minutes, however, he comes back to check "us" out again, showing now a shy respect for the water. He does not go into it, but plays at the edge pulling out some of the soggy leaves.

The water does not interest him for long today, and he flies onto a nearby stump. Suddenly, as if possessed, he pounces onto his moss hummock. He pounds one foot into it, then the other, with all the

Bubo and his reflection in the rock pool, when he is older.

force he can muster—one, two, three—harder, harder, harder. Then he bites into it, yanking up tufts of moss.

Within another minute he returns to his stump and serenely stares at me for at least ten minutes. Is he hatching up a plan? Apparently so. He flies over to me, and with his left foot clutches my shirttail while trying to hop away on the right foot. Alas, he very soon comes to the end of the shirttail, but he continues to yank it, determined to make off with it, not that he would do anything with it if he could get it. Nothing gives. Now he adds wing power, beating his wings violently. Still no progress. He becomes irritated, making the loud chirring noises that indicate his irritation (as when he is being held). He won't give up. Instead he changes tactics. He lets go with his left foot, turns around, and tugs at the shirt with his bill, bracing himself backward on both of his heels. More irritated chirring. Now he tries it with backward stroking wing-beats as well. Still no success. There are some things that can't be done in life, no matter how hard you try. But if Bubo tries, he tries hard. Perhaps he could never hope to be a successful predator without being persistent, even if the persistence is at times pointless.

Tocqueville said that we succeed at enterprises that demand the positive qualities we possess, but we excel in those that can also make use of our defects. By definition of evolution, *Bubo* has the best possible traits for survival within the constraints and possibilities that his *ancestors* have experienced. But that does not make this owl perfect now, unless it is within specifically defined constraints, such as the ability to resist uselessly expending energy by pulling on loose shirttails.

JUNE 16

Bubo looks at me as I come jogging up the path to the cabin carrying, as usual, a plastic bag with road kills. In an instant he has flown over beside me. With some relief I shake a rotting grackle from the bag. Despite its foul smell Bubo does not hesitate to grab it and fly off with it to a stump under the large spruce. Ninety minutes later he is still there, clutching the bird and occasionally nibbling at it. He has eaten very little of it, if anything, even though he has plucked out the wing and tail feathers. It seems he has reached an impasse—he is too greedy to let go of it, and too nauseated by its decay to eat. Undoubtedly anything he gets with feathers on it in the wild is fresh, so normally he does not have to reject any bird in his clutches.

Decayed or not, the grackle has feathers, and as prey it it worth defending. As I approach him, he puts his back toward me, snaps his bill threateningly, and extends his wings around the bird as if to keep me from seeing it or reaching it. All the while, he repeatedly grabs at it with his feet, as if to get a better hold on it. I leave him to it.

In the evening he exhumes the grackle from under some brush. But its quality has not improved on this warm day. Long after dark I hear something on the roof, and shining my flashlight up there, I see him on the ridgepole with the grackle still in his clutches.

June 17

Each time we go to our pool in the stream we throw a few rocks onto the dam. Pulling up rocks in the stream attracts minnows like plowing a field attracts seagulls. A scoop of the insect net near our feet usually captures two or three. Does Bubo like fresh fish? I wonder if he would catch fish out of a dishpan and eat them.

For at least 10 minutes, Bubo stares at the dozen minnows darting about in a pan half full of water. Then he approaches for a closer look. He meekly dips his bill in, as for a drink, squealing his bath noise, which sounds like the squeak of a stepped-on rubber duck. Then he pecks into the water and comes up with a minnow dangling from his bill. Crunch, crunch, and he swallows it head first. There are plenty more where that one came from. He hops into the pan, fluffs out his feathers as if to bathe, makes some more of his bathing squeaks, drinks a little, and then trots around and around, dunking for a minnow here and another one there. Six or seven mangled minnows are soon floating belly up. I do not think he is hungry today. Nevertheless, he makes use of the opportunity to play. Perhaps he is learning fishing techniques that could prove to be a useful skill in the future.

Catching minnows in a dishpan is easy for Bubo, perhaps too easy to be biologically meaningful in the wild. But this is only his first try. [In Asia there are owls that specialize in capturing and eating fish, and even owls in New England catch fish on occasion in the wild. One February, when most of the streams were still covered with ice, I found a barred owl dead alongside a highway in Vermont. The owl had ample fat. Its stomach contained two field mice and two central mud minnows, *Umbra limi*. These fish spawn in early spring in flooded areas adjacent to streams. I doubt that barred owls specialize on fish, though. Like great horned owls, they sometimes have to take anything they can get. Another barred owl that I picked up along a road in Vermon in mid-December even had the remains of a mole in its stomach.]

And while Bubo is sampling fresh-water food, I will reacquaint him with clams. Only this time I will prepare them a little better, the way *I* think they taste good. Maine clams fresh from the coast, whether steamed or fried, are just about my favorite food. Shouldn't fresh-water clams with salt on them be just as delicious as salt-water clams? After all, they *look* very similar, so why shouldn't they *taste* similar?

I collect a bucketful of clams from our stream, go home, and steam them. The shells open neatly and the meat falls out in nice firm packets. So far, so good. The odor, though, is not too inviting. I give Bubo a steamed clam, and he most definitely likes it: he swallows it in one gulp. So I decide to give it a try too. Although I am over fifty times his weight, I cannot swallow the large portions he can. I must try to chew mine into smaller chunks before I can hope to swallow them.

My clam has the consistency of shoe leather. It is difficult to imagine how something that looks like a blob of jellyfish can turn into a challenge to the molars after a short stay in boiling water. Perhaps I could chew it if it were crisped up a bit and made brittle. So, I bread up a few handfuls of clams and deep-fry them in corn oil. Now I am ready to make my second probe. The clam is crunchy, chewy, and indeed quite a bit more malleable. Only one problem: I taste something entirely different from before; the clam now tastes like the smell of the bog out of which the stream flows. Maybe I ruptured its stomach.

Clams are filter feeders. They trap particles floating downstream in a layer of slime and then move the slime with its trapped particles of bacteria and mud into their mouths. The mass is then digested in a stomach. If these clams end up tasting like mud, is it because they *contain* mud?

I slit open a plump juicy clam, and sure enough, the insides are bulging with greenish-brown ooze. So this is the bad-tasting stuff? I don't care to make a direct verification; indirect will do. I slit open a few more clams and wash them in a pan of clean water that soon looks like stirred-up bog water. I deep-fry the clams of this batch also. They are edible, though somewhat flinty after the long deep-fry. Would Bubo like to taste one? He plucks one from my hand and crunches it in his thick bill, sending the chipped chunks off in all directions. He crunches it a few more times, as he would an insect with a thick exoskeleton, and then he swallows it.

End of experiment. Now I know how Bubo likes his fresh-water clams: raw and camouflaged with hair, steamed, or deep-fried. None of these alternatives is the right one for me.

Attack Cues

ALTHOUGH Bunny delivers rodents, rabbits, birds, and shrews on a fairly regular basis, he is a fair-weather hunter, just like I am sometimes a fair-weather jogger. It was inevitable, therefore, that sooner or later the day would come when no dead animal is on hand. Bubo expects to be fed nevertheless, and when he wants his food he screeches incessantly in an irritating, rasping voice that becomes excruciatingly annoying. I have felt pleased so far to have been able to feed Bubo the already dead animals that would otherwise have gone to feed bacteria and maggots. Although I have nothing against insects, I would rather that a road kill provide Bubo with just one meal than have it spawn several thousand new flies. This is a purely emotional preference and has no logical justification. For better or worse, I place different values on different lives; for example, if I have to choose a live prey for Bubo's meal, I would choose a red squirrel or blue jay over a chickadee. Squirrels and jays are very common near Kaflunk, and both are significant consumers of the eggs and young of songbirds. On more than one occasion, I have seen a red squirrel emerge from a bird's nest with a naked chick in its yellow incisors. Jays are protected by law, but red squirrels are not, and squirrel meat is one of Bubo's favorites. If I were not here to act in his behalf, he would probably catch them on a regular basis himself.

I must now make my first conscious decision to choose Bubo's life over that of his prey. Bubo watches me as I leave the clearing with my .22 rifle, but I do not expect him to trail me because he has never followed me when I go for a jog or check the mailbox. Nevertheless, he seems to eye me suspiciously as I leave for the woods in a direction other than the one I usually take, down the trail. After I have walked for five minutes, he suddenly makes an appearance, flying down through a hole in the canopy of the spruces. He lands beside me, and as I continue to stalk "our" prey, Bubo runs on foot behind me. He keeps falling farther behind, but when I am almost out of sight, he flies and lands ahead of me again. I find it very strange that he is following me now, because he has never done so before. But I do not mind: he is now no longer screeching. He is totally silent. That is also very strange.

On one of my stops to look and listen for red squirrels, Bubo is no longer with me. Behind me I hear a squirrel, and turning back I find Bubo already there. He is watching the squirrel up in a spruce tree. The little red bundle is chucking and scolding with its tail plume upright, sitting on a branch close to the trunk, stamping its feet up and down. Bubo has seen red squirrels near the cabin many times before, and he has never made an attempt to chase one. As before, though, he watches the ebullient rodent with interest. Might he attack one now? I anxiously wait, but he shows no intention of giving chase. I am a little disappointed in him, and finally I have to shoot it for him. But my shot is not good.

The squirrel tumbles to the ground, wounded. I fear that it might escape and suffer needlessly. But will it? Bubo dives through the branches, breaking and scattering twigs in his haste as the squirrel tries to scamper away. There is nothing wrong with Bubo's attack technique, and in seconds the squirrel is securely in his talons. I am amazed at his resolve, and success, realizing only in retrospect that it would have been totally unreasonable for him to try to outmaneuver a healthy, alert squirrel secure in a tree. The amazing thing is that he did not learn to capture this squirrel from direct experience; he acted as though he knew exactly what to do, and when to do it.

For the next ten minutes Bubo continues to subject the squirrel to the strong clenching movements of his taloned toes. The movements are so quick and vigorous that it looks like all of his sharp spikes are stabbing in unison. His feet are now indeed his lethal weapons. Not until Bubo's foot-clenching has squeezed the squirrel senseless does he also bite into it repeatedly, starting at the head and neck. He eats half the squirrel on the spot, caching the remainder on the ground under a small fir tree and staying afterward nearby up in a spruce.

June 20

As I jog down the trail past Bubo sitting on his favorite birch, he looks at me briefly and continues to preen himself. He has no intention of following me. Further on down the trail, I find a dead mouse that

Bunny had caught but neglected to deposit under our bed last night. I hold it behind my back and call out, "Bubo!" He instantly turns around on his perch, looks at me for a second, and, swooping down the trail close to the ground under the pine branches, lands beside me.

Bubo follows me because he expects me to have a constant supply of mice on hand. I now often take one or two with me. I am reminded of Merlin the Magician, who carried dead mice for his owl Archimedes under his hat, where it was "most convenient"; but I usually carry mine in my hip pocket. Also, Bubo, unlike Archimedes, does not ride on my shoulder. Perhaps Merlin would have preferred it that way, too: unlike Merlin, I seldom have white streaks down my back.

June 27

This morning Bubo seems more hungry than usual. For about two weeks now, his begging noises have sounded like blasts of escaping steam or hoarse screams. He makes them only when he is near me asking to be fed. It seems rather late for him still to be carrying on this way, but then, who knows about owls. As with a baby's cry, I find the noise distracting. There is only one way to shut him up: give him food. Today it requires a mouse, half of a young rabbit, one warbler, and one automobile-flattened young ruffed grouse. Perhaps adult owls react as I do and frantically get some food to stop all the racket.

It is now easy for me to locate Bubo. His grating call serves to remind me of his presence: if he knows I am nearby, I cannot overlook or forget him.

July 1

As evening approaches, Bubo sits in front of the cabin, looking in at us through the window. His begging calls are incessant and annoying, but I am trying to ignore them: if he is not hungry, he might never be motivated to hunt on his own. Perhaps I can give him a pacifier. I throw out a banana peel. He pounces on it immediately and tears it to shreds, while occasionally shaking his head in disgust or irritation. He does not eat any of it.

Bullfrogs, one of the known prey of great horned owls, live plentifully in a local pond. Perhaps some of these amphibians could be of help in sharpening Bubo's hunting skills. I call five unwilling volunteers into service and present them to Bubo. I release one of them; Bubo stares at it and makes one hop toward it while the frog makes one hop away. Bubo's next hop, however, is right on target. His bill crunches into the frog's head, and then he eats it, though with a certain lack of enthusiasm. He swallows mice and other delectables in one gulp—in seconds—but Bubo takes fifteen minutes to consume the frog, tearing it up bit by bit, slowly and dispassionately, like a child forced to eat spinach.

Bubo calls for food again in the evening. I release bullfrog number 2. This individual does not hop away when Bubo confronts it. Instead, it blows itself up and rises off the ground on stiffened legs to make itself appear bigger and menacing. But Bubo is not intimidated. He walks right up to it, moving his head this way and that to get a closer look at the odd display of the beautiful creature. He leaves it alone. Beauty is only skin deep.

July 2

Bubo is even more hungry today. I offer bullfrog number 3, which promptly hops away. Bubo shows no interest in chasing it. Maybe he needs to see blood, so I chop up the frog. Is this better? He limply takes the meat into his bill but seems to decide this is disgusting stuff and drops it immediately. I try again with a skinned frog leg, whose meat to my eyes looks like the white meat of chicken. Whoever said frog legs taste like chicken has not consulted an owl. No doubt, Bubo would like them fried.

I play on Bubo's apparent preferences for gourmet foods. I take the skin of a red squirrel and put a bullfrog inside it. *Grenouille en écureuil* for Bubo. He recognizes the squirrel from its fur, and in one lunge grabs it from my hand, loudly chirring his enthusiasm. He even defends his meal with loud bill-snapping when I get close to him. He pulls out the frog meat, tears it up, and eats it with the gusto of one who had never tasted anything better. Apparently he relishes the frog meat not on the basis of what he actually tastes, but on the basis of what he has been led to believe it is.

There was nothing wrong with the meat I gave to Bubo, yet one might almost think he refused to eat it for the same reason that, say, Hindus refuse to eat beef. We humans invest a great deal of energy in ignorance when we feel a need for illusion. But I thought that *he* would value objective facts above illusion or belief. Still, his eating of the frog meat, when he *thought* it was squirrel meat, clearly shows that he is no more objective than many humans.

I offer him another frog, this one disguised in the skin of a cedar waxwing. Two long, naked frog legs protrude conspicuously from the feathers. This is one suspicious-looking bird. Bubo takes it anyway, but instead of swallowing it whole, as is his usual custom with small birds, he tears off only little pieces of meat. He finally stops trying to eat more and caches the rest of my offering in dense grass. Perhaps he has finally caught on. There is, after all, a limit to his gullibility.

I am puzzled why Bubo finds the meat of bullfrogs so distasteful. Meat is meat, I had thought, and great horned owls in the wild are known to catch almost anything that crawls, hops, or flies. And that includes bullfrogs.

July 3

It is raining. After a long rain Bubo usually sits quietly, pulling his wings tightly against his body like an overcoat. But today, despite the rain, he perches out in the open, fluffs out his feathers, and opens his wings wide and shakes them while hopping excitedly. He is either doing a fair rendition of a rain dance, or he is taking a shower.

July 5

I found no road kills yesterday, so this morning I go out on my second red-squirrel hunting expedition for Bubo. As on the first trip, Bubo seems to sense something is up when I go off north into the woods, instead of south down the wooded path, and he comes along. Within ten minutes a squirrel appears from its hiding place in a dense spruce and chatters at Bubo. Together we capture it, much like we had captured the one before. He again flew after it as soon as it came tumbling down into a dense fir thicket, where it was out of our view. Nevertheless, Bubo hunted for it, and found it.

The squirrel's chattering may be a signal to a predator that chase is futile because the animal is alert. It saves the squirrel the trouble of having to flee. Only now, since I had a rifle, this previously successful strategy was the rodent's downfall. The same old script does not always work; animals are perhaps only seldom, if ever, "perfectly" adapted, because evolution never stops. Each species carries evolutionary baggage from past ages, and it requires strong selective pressure to unload it.

JULY 7

Yesterday I brought back two chipmunks. They seem to be the most common road kill on the local highways this year. As I come trotting up the path to the cabin, Bubo starts to screech as soon as he sees me, just to let me know he is around. If I call "Bubo" he launches himself into flight instantly and lands at my feet. I feel like an owl mother returning back to her nest area, and my owlet wastes no time in trying to relieve me of my catch.

So far Bubo has been undiscrimating in what he accepts for food, except for the frogs. He accepts a bright yellow gold finch as readily as a scarlet tanager or a black-and-white chickadee. Can he distinguish among colors at all?

NOCTURNAL animals generally do not have color vision, which depends generally on three populations of sensory cells called "cones," each maximally sensitive to three different wavelengths of light as determined by the color of the pigment they contain. The color the animal sees depends on the relative stimulation of these different cones. The rods, the sensory cells responsible for black-and-white vision, respond maximally to blue-green light, so objects of this color seem unusually bright in dim light. The eye of the great horned owl, like the human eye, contains both rods and cones (Fite, 1973). Evolutionary evidence also suggests that owls may have color vision.

Bunny brings in a variety of small nocturnal mammals. These come in different colors. The jumping mice are a straw yellow, like the dead grass in the clearing where they are found. The woodland voles have a dark red-brown back, like the dead leaves on the forest floor where they live. The deer mice are grayish brown. These colors serve as camouflage. When wild mice of natural colors were placed in enclosures of different soil, owls preferentially captured those that did not color-match their background (Dice, 1947; Kaufman, 1974). The color of the mice is thus said to be "adaptive," and if this is true, then owls and other predators have been responsible for its evolution.

We do not know if the same holds true for all owls, but behavioral experiments show directly that the nocturnal tawny owl, *Strix aluco*, has color vision, although it is difficult to train these birds to use color discrimination (Martin, 1974). This owl has cones with oil droplets containing pigments that maximally absorb yellow, green, and blue (Bowmaker and Martin, 1978), much like the trichromatic color-vision systems in humans and other vertebrates. In addition, there is a fourth type of cone with red absorbing pigment (from 400 to 650 nm). However, the cones with red pigment, also found in pigeons, are rare—only one percent of the cone population. Indeed, the cones for color vision constitute 10 to 20 percent of the population of sensory receptors, while in diurnal species they constitute 70 to 80 percent. The tawny owl is thought to have greater visual acuity than the pigeon because rods cover a greater area of its retina and because there is a greater density of these receptors per area. Bowmaker and Martin (1978) conclude: ". . . it seems reasonable to suppose the owl's photic system is the remnant of a more highly developed ancestral system (for color vision) similar to that in, for example, the pigeon." In a sense, then, color vision in owls may be evolutionary baggage that is now of little use to them, although the selective pressure to get rid of it may not have been great enough to dispense with entirely.

I have been a little suspicious of Bubo's purported superior night vision since he banged into the rafters while flying one night in the cabin. Also, since on an overcast day Bubo already opens his pupils to three-fourths or more, the camera equivalent of f/5 or f/6, it seems that his visual ISO is not all that remarkable.

Different species of owls require different minimum illumination to see, and these light sensitivities in turn affect when the owls are active (Erkert, 1967, 1969). Some nocturnal owls were thought to find dead prey visually at light intensities approaching down to 0.00000073 foot candles (Dice, 1945). But light intensities at night in a shaded forest are even lower than this, and an owl's vision is not always enough to capture prey. Indeed, more recent studies (Martin, 1977, 1978) indicate that the eye of one nocturnal owl, *Strix aluco*, has a visual sensitivity that is very similar to that of the human eye. Hearing therefore assumes primary importance in many species of owls.

Nevertheless, I wonder if they can see the infrared radiation that is emitted by all warm bodies. If an owl, like a pit viper, had an infrared detector, then it would have a wonderful tool for catching birds and mice at night.

IT would be nice to give Bubo an eye test. There might be a way. But first he must be hungry.

After two days without food, Bubo is ready to listen attentively and to look. It is an overcast night. I toss a dead deer mouse onto the ground near him. When it lands white belly up, it is visible to me, though only as a faint white blur. Bubo makes soft grunting sounds, and in a second or so he pounces on the mouse. So far so good. When I toss another mouse and it lands belly down, it is another story. Bubo must have heard it land, because after a few seconds he hops down near it. He hunts for it, but he cannot find it. Several trials later the results are the same: he cannot see dead, brown deer mice on the forest floor directly in front of him at night, but he has no trouble seeing white mice.

The early literature (Vanderplank, 1934) had suggested that tawny owls are sensitive to the infrared emitted by living warm mice. But several subsequent studies have suggested that this notion is wrong (Hecht and Pirenne, 1940; Dice, 1945; Hocking and Mitchell, 1961).

To find out if Bubo can see the brown mice when they are in the dark, but warm, I heat one in the hot air of a double boiler and test him again. He behaves exactly as before: he knows I am tossing food in front of him and he looks for it, but he does not see brown mice even when they are emitting infrared radiation. All in all, I conclude from my "quick and dirty" experiments that Bubo's abilities to see at night or to detect infrared radiation are not impressive. Most likely, he can see a mouse better in the dark than his potential competitors, such as the diurnal birds of prey, and it is the relative difference from them that counts, not his absolute capacity.

Sound is a major dimension in an owl's environment; it is of far more importance than it is to humans. An owl, with its extremely acute auditory sense, may have to sort out the sounds of a beetle scuttling in dry leaves, a deer walking, a rabbit hopping, a gopher burrowing, and a porcupine ambling through the underbush from that of a mouse running. To catch the appropriate prey it must also differentiate these sounds from the background noises of leaves rustling in the wind, falling twigs, and scraping branches. To be able to strike their prey through sound detection alone, as some owls do, the birds must not only localize the sound but also identify it accurately. Barn owls are able to memorize amazingly fine details of sound, when those details help to distinguish a meal (Konishi and Kenuk, 1975). But just how important sound is for prey location and identification in great horned owls is not known.

JULY 11

Bubo's new breast feathers are about half in now, and on each side of the top of his head two dark shoots are sprouting through the fuzz. They are the first hint of his "horns," or "ears," that will identify him as an adult. Each will eventually consist of about a dozen black-tipped feathers. His wing coverts are now smooth and shiny. They overlap each other like shingles and produce a silky surface that is now nearly impenetrable to water. The outlines of the individual feathers themselves are not visible, as they are on a crow, for example. Only the primaries and secondaries used for flight can be seen as individually distinct.

It gives me a thrill to see the big bird come flying gracefully over the birches, wheeling down into the clearing, and landing beside me when I call him. I will soon be able to lead him to different birds in the forest to test their mobbing reactions and, in turn, to observe *his* reactions.

His wing beats are usually shallow and rapid. Before he alights on a large branch, he often approaches it from underneath and swoops up, braking himself with stationary outstretched wings, and landing daintily. Sometimes, however, he beats his wings so rapidly as he makes his landing that it looks as if they are vibrating. At other times he is more bold. He flies directly *at* a branch without trying to slow his flight. Because of the absence of aerodynamic braking, he is brought to an abrupt halt by the force of his striking the branch. It looks like a variation of his "attack" game.

JULY 15

I release a toad in front of Bubo. It manages to make only two or three hops before Bubo jumps onto it and grasps it firmly in the talons of his right foot. Having pinned his prey, he looks around in all directions, as he always does before proceeding further. The toad squirms and then stops moving. After Bubo is satisfied that there are no potential interlopers to his gourmandizing, he bends down with closed eyes and starts to nibble very gently at the toad. But then he lifts his head, as if in surprise, opens his eyes wide, and violently shakes his head in the manner of one who wants to emphasize "No, No, No!"

After these vigorous gesticulations, he rises to his full height, pulls his head back for a fuller view, and peers at the toad again, more carefully this time. Gradually the talons of his foot retract, releasing their grip. Then he walks slowly backward, still peering at the toad while shaking his head at intervals. Meanwhile, the toad hops away, acting no worse for the wear. Toads are covered with noxious secretions (although their meat might be sweet). Maybe he learned something. And after you've tried to eat a toad for breakfast, nothing much worse can happen to you. Maybe.

I have a handful of turtle meat that had been in the warm air for a couple of days, and it smelled accordingly. Although it is nauseous to me, Bubo, proving he is indeed hungry, grabs it eagerly and starts tearing it apart and swallowing it in chunks.

The Craighead brothers (1969), in their studies of this owl in southern Michigan, found that the diet of the local great horned owl consisted of 40 percent mammals and 60 percent birds in the winter of 1942, and 62 percent mammals and 30 percent birds in the winter of 1948. The Craigheads concluded: "The most distinctive feature of the great horned owl diet, as compared with that of other raptors, is the wide range of prey species that it includes." That wide range of prey apparently also includes decayed turtles, but not live toads.

According to Arthur Cleveland Bent (1961), in his classic *Life Histories of North American Birds of Prey*, part 2, the great horned owl feeds on "almost any living creature that walks, crawls, or flies or swims except the larger mammals" (which does not exclude "large and small skunks, woodchucks, porcupines and domestic cats"). "Living" seems to be the common denominator for the great horned owl's diet, although most of the food Bubo gets from me is long since quite dead. What would he do with a moving mechanical toy that mimics live prey but is not edible?

I look for a mechanical mouse at the toy store in town but have no success. I settle for a little red plastic bear and a green tin bullfrog. The bear, after being wound up, moves forward in little circles, rocking back and forth, all the while buzzing like a cicada. Bubo watches it in apparent fascination, cocking his head and blinking his pupils. There are only a few things that a great horned owl will not attack, and a buzzing red bear seems to be one of them—at least on this occasion.

The tin frog is another matter. By some strange transformation of energy the wound-up green frog scuttles forward in a frenetic series of little hops. Bubo leans forward and looks, and then flies down from his perch in the birch for a closer look at this thing on the plywood platform below him that acts and looks a little like a familiar, though not so tasty, animal. He waddles up to it and without further ado he attempts to get a hold on the unyielding carapace with his hooked bill. Not much success there. He pulls his head back, takes another look: this is *not* an animal! He leaves. He has solved the mystery and it ceases to be of further interest to him.

Having now gotten into the spirit of the thing, I continue the game with a slightly new twist. I tie a patch of rabbit fur onto the head of the bear; then I wind the bear up to let it go as before. I visualize what will happen next: little bear will spin and buzz off across his plywood arena; Bubo will hunch his head forward, lower his second foot from out of his breast feathers onto the limb, and then push off, head first, for a silent glide to the bear; in the last moment he will lift his head, throw his feet forward, and drive his extended talons in. But this scenario does not occur. Bubo does not pounce. He only watches with mild interest. The bear, even in its new rabbit-fur coat, elicits no response. Maybe Bubo has become wise to my ways, or maybe he knows what a furred animal *really* looks and acts like: this buzzing and tinny thing is surely a fake. More experiments are in order, and for the time being I remove the rabbit-fur accoutrements from the toy. Bubo jumps down, waddles over to them, and eats them.

July 16

This morning, while he is perched on my hand, Bubo hoots again for the first time since he made those tiny whispers when he was still a nestling. I answer him, he answers back, and we answer each other back and forth six to eight times. Not only do we hunt together, we also sing together.

Companionship

We have begun building our log cabin in the clearing by the old Adams farm site below Kaflunk. We will be close to the old well, which we have repaired. Also, Kaflunk is small and crowded. Next year it will be even more crowded: Margaret and I are expecting a child. It is time to think about roomier quarters. I am energized by the prospect of the child, and the work becomes easy. And besides, it also gives us pleasure to look around at the end of the day, or year, and actually to see what we have accomplished.

Last summer, in my spare time, I chopped down one hundred spruces and firs in the surrounding woods. Margaret helped peel the bark off the freshly felled trunks. Mert Farrington came up with his pair of brown oxen, Chub and Toby, and in two days all the logs were in a pile in the clearing. This summer we hope to have the walls up and a roof over them. Although so far we have notched and put in place only twenty logs, we already call it our log "cabin." Wishful thinking. Bubo often comes to investigate. He uses the cabin as a perch. Does he also come just to be with us? In my selfish way I hope so, because I am getting attached to this bird. He is becoming part of my life.

July 18

We see Bubo on his birch at Kaflunk as we go down the trail through the woods to the log cabin. He does not follow. Nevertheless, a few minutes after we reach the cabin he comes wheeling in from out over the forest, and he lands beside me and hoots. Did he know where we were going?

Two days ago, he inexplicably started to hoot loudly whenever he has met me outside the direct home area of Kaflunk. This is something new. I take it as a friendly greeting. Having said his hello's to me, he hops over to Margaret, gives her a hoot, too, and then flies back into the woods. He begs for food with his grating screech only at Kaflunk.

We do not see much of him these days. He comes once a day to get fed at Kaflunk and to make a social call when we work at the log cabin. Perhaps he is just snoozing on some shady branch deep in the woods, or he might be out practicing to hunt on his own.

BEING a good hunter may be especially important in the life of a male owl. Throughout courtship, during the month that the female incubates the eggs, and during the weeks after that when she stays at the nest day and night to protect the young and keep them warm, it is the male who is the sole provider. There are probably great differences between individual males in their hunting ability and in the quality of the territory they hold. Before the females of many species commit themselves to raising a family with a male in any one territory, they first assess his ability as a provider and the quality of the territory he is able to secure and defend. That mate assessment in owls occurs in courtship is suggested by the following observations of great horned owls by Floyd Bralliar (1922):

> So he began bowing his head, ruffled his feathers, raising his wings and spreading his wings in a curious manner. . . . Aside from watching his antics, she took no notice of his presence. Growing more earnest, he began hopping from branch to branch, continuing his maneuvers and snapping his bill fiercely as if to show that even though he was not as large as she, what he lacked in size he made up in bravery.
>
> Finally, he attempted to approach and caress her but she ruffled her feathers and rebuked him sharply. He took flight, sailing up and down, around and around, evidently doing all the stunts of his race, now and again punctuating his efforts by snapping his bill. After a few moments he alighted again and began his bowing and dancing all over again.
>
> A rabbit came running down the bank and its white flag caught his eye. Rising in noiseless flight, he sailed downward without the flap of a wing, caught his prey from the ground, glided back into the tree, and presented his offering to his lady love. Apparently, she was convinced of his sincerity. Together they devoured the rabbit, and when he again began his love dance she joined in with as much enthusiasm as he.

Was it "sincerity" that swayed her, or a meal and proof of his ability as a provider?

JULY 19

This morning Bubo arrives five minutes after I call him, takes the whole squirrel I give him, and carries it into the woods. A half hour later he returns, making his hoarse, squawking, begging call. He could not possibly be hungry. Does he want food to hoard, or is it something else?

I give him a jumping mouse. I know he cannot resist it, but he does not swallow it immediately. Instead, he spends a lot of time crunching its skull and other bones. Then he perches on my arm, mouse in bill, procrastinating. A jet passes over so high that it is barely visible to me. But he watches it the entire time it takes to cross the horizon, the mouse all the while dangling limply from his bill. He makes some muffled croaking noises from deep within his throat, and then, when the aircraft is gone, he returns his attention to the mouse. He tears off bits and pieces, from the head backward, and eats one tiny morsel at a time.

Having finished eating the mouse, he hops to the top of the open screen door to examine Bunny, who is below him on the threshold, taking a nap in the morning sun. Bunny gets up and continues his nap elsewhere.

Bubo, filled to the brim with his favorite food, is getting warmed up this bright and sunny morning. He hops down to his moss hummock and stands there a few seconds, solemn and dignified. In the next instance, he strikes with his talons into the moss at a frenzied rate of about five times per second. It is more than he can manage, and he collapses on his side and gets entangled in a loose rope: but he keeps on striking, tossing billfuls and fistfuls of moss in all directions. Now his attention shifts from the moss to the rope. He pounces first at one of its tangles with his right foot, and then at another with his left foot, flapping his wings to keep his balance somewhat. Then, as suddenly as he had begun his attack, he flies off with quiet and graceful wing beats.

JULY 23

I meet Bubo deep in the woods today, near the brook. He comes to greet me, not to beg as at Kaflunk, but with a soft hoot. It is like a handshake, and it dispels that slight veil of uncertainty: "Yes, this *is* Bubo, and I recognize you."

He watches me walk back up the path but does not follow, and he does not appear at Kaflunk in the evening even though I call him repeatedly.

Bubo is becoming independent. We exchange warm pleasantries when we meet. They are genuine, not mere courtesies. But he needs me less. I want him to become independent, but still I miss him when he is out there, somewhere. I do not know whether to give in when he begs or to let him stay very hungry. To be safe, I feed him, although it is likely that his parents would have left him by now.

DESPITE their great solicitude, owl parents do not pamper their young for as long as they demand food. The young learn to hunt because they ultimately have no choice. Perhaps Bubo has not succeeded unaided in the hunt for squirrels and rabbits so far, not because he does not have the capacity to succeed, but because he has not been hungry enough to develop his capacity.

Paul Errington (1932) in his studies of the great horned owl in Wisconsin in the early 1930s, tethered young near their nests in order to study the birds' food habits and feeding patterns. He found that by the middle of July the parental devotion toward the tethered young was decreasing, regardless of whether or not the young were still constantly calling for food:

> The length of the feeding period of young by adults does not seem correlated with the variable individual needs of the young as affected by circumstances. The 'weaning' of mid-summer culminates the protracted juvenile education, whether, it may be presumed, the young have learned to hunt or not. I am convinced that no normal juvenile horned owl fails to take advantage of family support as long as such is to be had. When deprived, it does what it has to do.

Errington (1944) later described a practical application of the methods he used to study the owl's food habits. He suggested they might be used in survival tactics of airmen lost in the woods. He says:

> One of the best bets in forested North American wilderness would be to mooch off the horned owls. This suggestion is not crackpot. Horned owls are among the most efficient of wild hunters, and, in the most heavily populated areas of their range, their numbers reach a breeding pair per square mile. . . . Quantities of food brought

daily to the young may be considerable and may be further increased by a few tricks on your part.

The "tricks" Errington refers to are necessary to guard oneself against attack at the owls' nests, since one can "compare a blow on face or neck by a horned owl with a blow by spiked brass knuckles," and "you can lose an eye or even suffer fatal wounds through carelessness." Having warned his readers, he proposes that each of the young be penned separately in cages made out of wood at hand. The cages should be "open enough to allow visibility and ready feeding by the owl or by yourself. Carcasses or pieces of prey animals left about the pens (by the adult owls) may be removed for your own use."

If the adults are forced to provide for several mouths, they are stimulated to bring more food. In Alberta, tethering sites with one caged young were provided, on the average, with 293 grams of prey per day (about one-half pound). When four young were being fed, the adults brought in 1336 grams per day (McInvaille and Keith, 1974), or 4.5 times as much.

July 25

Bubo's ear tufts or "horns" are nearly grown to their full length. He now truly looks like the great horned owl clothed in adultlike plumage. Unlike some other owls, the juveniles of this owl have nearly the same plumage as the adults.

I do not know the actual function of his ear tufts. However, he has a fondness for perching on broken-off tree trunks, where he quite effectively impersonates the top of the stub; his ear tufts look like ragged edges that help him blend in, so that prey (or predators?) may be less likely to notice him.

SUBSEQUENT to my speculation, I have found in the literature that the adaptive value of the ear tufts has also interested others. Ear tufts are found not only in the great horned owl, but on 50 of the world's other 131 species of owls. It has been proposed that the tufts mimic the ears of mammals, aiding in threat displays (Mysterud and Dunker, 1979); as short-range species-recognition marks (Burton, 1973); and as camouflage (Sparks and Soper, 1970). Perrone (1981), who has reviewed the incidence of ear tufts, reports that they occur in none of

Bubo when older, with
fully-grown ear-tufts.

the twenty-one species of diurnal owls; among the nocturnal forest-dwelling species, forty-nine species have conspicuous ear tufts and sixty species do not. (*Bubo* was classed among the nocturnal, forest-dwelling species.) Examining the ecological evidence for the three hypotheses, Perrone concludes that camouflage is the most likely function.

WE HAVE not seen Bubo for almost two days, except for one quick social call at the log cabin. However, today he shows up at Kaflunk, where he greets me with his grating begging scream as soon as I step outside the door. Here at home he rarely gives his soft, reedy "hello" hoot. None is necessary. At home he expects food, not the exchange of greetings.

He plays again with a wood chip and the rope, and he attacks the moss hummock. The way he jostles on top of that thing reminds me of a drugstore cowbody riding a mechanical horse. He never seems to tire of it.

I had begun to think that other birds pay little attention to Bubo, but today I am proven wrong. Bubo is perching quietly, minding his own business, and a flock of about ten chickadees is nearby, and many of them are giving warning calls. Several come to within one to two meters of him. A black-throated green warbler arrives and also gives a few excited "chips." Most of the birds are young. Most likely none of the youngsters has ever had contact with an owl before.

Each situation that arises gives more clues about why birds mob owls, even though I have no idea yet how the clues will ultimately fit together. Each set of clues also brings more questions. That adds to the fun. Discrepancies from the expected are exciting because they show that there is something new, or that something has been overlooked.

JULY 26

Bubo must be hungry this morning, and I have a toad waiting for him to do the toad experiment again.

The toad hops in front of Bubo, who pounces on it after a somewhat cursory inspection, lets it go, pounces on it again, and nuzzles it lightly with his bill. After several pounces there is not a scratch on the toad, and it leisurely scrambles away over the moss.

Bubo, not very camouflaged, perched on Kaflunk.

I think he was just playing with it. He did not show any surprise, and he never shook his head in disgust as he had with the first one. I don't think that the idea of eating this creature occurred to him. To him it was just a toy, like a wood chip or a piece of rope that I might pull along on the ground. However, if I had not known the details of his previous behavior, I could easily have mistaken this encounter as indicating a lack of knowledge of toads. But there was little doubt—he knew very well what toads are good for, and what they are not good for.

I want to get a photograph of Bubo defending his food from the cat. To prepare for that I tie a dead squirrel onto a heavy rock with several strands of fish line. I get the camera ready, and while he is picking on the squirrel I call to Margaret to bring the cat. She does. In one big heave Bubo is gone, with the squirrel in his talons and the fish line trailing behind.

He does not visit us at the log cabin in the afternoon, nor at Kaflunk in the evening.

July 27

I wake up during the night with a jerk. A thought strikes me like a hammer: Could Bubo have gotten entangled in the fishing line and strangled himself? Several years earlier a fisherman found a dead great horned owl at Bog Stream nearby, entangled by just such a fishing line. I wondered then how it could have happened. But knowing Bubo's ways, I suspect that the unfortunate owl was attracted to the string or the red and white float on the line, because it was something "new" and different in the environment to investigate.

I tumble out of the cabin and call him. After four desperate, loud yells—what a relief—the familiar dark shape comes gliding over the forest.

July 28

Now, in midsummer, the behavior of the birds has changed dramatically. Flocks are gathering. A loose foraging aggregation of chickadees, nuthatches, warblers, scarlet tanagers, juncoes, and solitary vireos comes by through the birches around Kaflunk, as they had months earlier in the spring. Blue jays and robins are here often, feeding on the blueberries. But the birds pay little attention to Bubo. They simply go about their business, even though he is in plain view. Physically, Bubo is indistinguishable from an adult, and I can no longer rationalize that the birds do not mob him because he is "only a baby."

One blue jay scolds for a minute or so, and so does a chickadee. Strange. Why do they act differently from the rest? Their behavior could easily be dismissed as just a minor incident because it is not consistent with my other observations and expectations. But *every* incident is of some significance. Each one has an explanation. Details that now seem trivial may later turn out to be critical notches in the key to understanding. Anecdotal observations should therefore not be minimized. If I can understand the reasons why one flock mobs the owl while others, at the same time and place, do not, then I will have gained a fundamental insight into mobbing behavior from which to build a more general hypothesis.

We leave Kaflunk, and Bubo and the flock of birds, to go down to the log cabin. When we get there he is already there, perched on the timbers, hooting a hello to us. He had left after we left, even though he had not yet dried off from the bath he had taken while we were still there. He must have experienced difficulty flying while waterlogged. He had interrupted his routine, just to visit us. Why? To say it is for food would be the simplest explanation. But it is probably the wrong one since he never begs at the log cabin.

It ALMOST seems that Bubo wants our companionship. But owls, particularly great horned owls, are not noted for their sociability. They are the quintessential birds of solitude. Then again, owls are not totally without social graces, especially when it comes to courtship. Great horned owls mate for life, but every year they become amorous all over again. Could Bubo have his eye on one of us as a potential companion or mate? If so, how would he let us know? Charles E. Bendire (1892) reports:

> I once had the good fortune to steal unnoticed upon a pair of birds (great horned owls) in their love making. The ceremony had evidently been in progress some time. When discovered, the male was carefully approaching the female, which stood on a branch, and she half turned away like a timid girl. He then fondly stroked his mate with his bill, bowed solemnly, touched or rubbed her bill with his,

bowed again, sidled into a new position from time to time, and continued his caress. All these attentions were apparently bashfully received by the female. From thereafter the pair flew slowly away side by side.

Walker (1974) reports on a pair of displaying great horned owls where the female "nibbled at the male's beak." The female then lowered her head and "the male ran his beak through her feathers in a touching caress, the epitome of gentleness."

The preening of another individual is technically called allopreening, where "allo" comes from the Latin word for "other." The function of allopreening in birds is somewhat of a puzzle. Monkeys do it to socialize and to pick parasites off each othe. But why do at least ten species of owls do it? The question was asked by Forsman and Wright (1979) regarding spotted owls, *Strix occidentalis*. They report:

> Our observations of 127 Spotted Owls between 1970 and 1978 revealed that most, if not all, paired individuals allopreened regularly with their mates during the spring and summer period. . . . Usually, the bird that initiated allopreening indicated its intent by staring at the other bird and uttering low cooing or whistling calls. If the bird was receptive, it usually stared back, sometimes giving low cooing calls. After this brief solicitation exchange, one bird would fly or walk to a position beside the other (if it was not already in this position), where it would lean out and begin to preen the other's head. Typically, allopreening birds perched side by side, facing in the same direction. Both birds partially or entirely closed their eyelids and nictitating membranes. . . . The recipient usually moved its head, as if to facilitate preening in whatever area was being preened. . . . One bird would preen the other for a period, the roles would be reversed, often several times in a bout. . . . While allopreening, owls frequently made vocal cooing or whistling sounds that were just barely audible.

Forsman and Wright found that allopreening in owls, although it has probably evolved into a pair-bond maintenance display (possibly derived from ectoparasite-removal behavior), was infrequent during pair formation, nest-site selection, incubation, or copulation. It occurred most frequently after the young were fledged and the adults again had time to themselves: "In July and August, allopreening

seemed to be an almost daily activity for pairs that roosted together."
They noted, however, that not all pairs habitually roosted together.
Some were less intimate: "The allopreening behavior of a female
Spotted Owl that we held in captivity between 1970 and 1978 appeared identical with that of wild birds except that, because the bird
was imprinted on Forsman, she directed her allopreening behavior toward him. She would not allopreen with strangers; she attacked
them."

Bubo's behavior with me is comparable. His nibbling of me is obviously not simply a body search for food. Forsman and Wright think
that the owl's "affection" may be related to their power: "In owls, the
gentle 'nibbling' associated with allopreening could be ritualized biting behavior that has become so modified that all appearances of
aggression have been eliminated. In raptors such a high degree of rit-
ualization is to be expected, because the weapons of aggression are so
well developed that even the slightest amount of overt aggression in
this type of display could lead to injury."

JULY 29

We went to Logan Airport in Boston to pick up my daughter Erica, and
when the three of us happily stumble up the steep trail back to Kaflunk it is 2:30 A.M. It is drizzling, and the dense pines block out the
remaining skylight. As we approach Kaflunk, we hear Bubo's begging,
steam-whistle scream from the darkness. So—he can be active at
night after all! I call him and he comes out of the forest, crash-landing
(not very gracefully, from the sound) in a spruce by the camp. Although he may be able to see in faint light, I doubt that he sees very
well in this darkness.

On this and on subsequent days he pays almost no attention to Erica. What a difference from the way he had reacted to Ken earlier!
Does he dislike only other men?

A Goodbye

WE HAVE to leave Kaflunk soon, and we will have to make a painful decision about Bubo. I would like to leave him here, giving him the freedom of the wild. But I am afraid that he might starve. To find out if his dependence is declining, I have kept daily records of what he has received from me. He refused food only rarely. Possibly he occasionally made a kill. But on the whole he probably still depends on me. It was his easiest, most efficient strategy. Why *should* he opt for a more difficult one? Perhaps I have not been harsh enough with him. A growing young songbird will get very hungry in a few hours, and may starve to death in a day if not fed. Predators are different. Perhaps I should have witheld food from Bubo for as long as a week or so, to make him become keenly motivated to hunt.

AUGUST 2

The white-throated sparrows have a nest next to the log cabin, with half-grown young huddling inside. Bubo, who perches daily above them, has not yet found the artfully concealed young. The parents, who ignored Bubo while they had eggs, now scold him incessantly with monotonous chirping as they perch almost motionless above him. By making themselves look slow and less agile than they really are, do they divert his attention from the nest?

AUGUST 11

Early evening. I have not seen Bubo for over a day. But within a minute or so after I come back inside after calling him I hear his familiar thump on the roof above the patter of the rain drops.

Now he is perched opposite my window in the big white birch. His left foot is completely pulled up and tucked deep into his sleek, hori-

View out the window of Kaflunk to Bubo's birches. Bubo is perched on lower left limb of birch on left side.

zontally striped breast feathers. His head, tucked down into his shoulders, sprouts two long tufts that stick out to the sides. Slowly, very slowly he rotates his head to the right, then to the left, as he peers at me lazily through partially closed eyes. First he blinks one eye, then the other. Once in awhile he cocks his head slightly to the side, rotates it a bit, thrusts it forward, blinks both eyes. He thrusts his face forward 2 to 3 centimeters and then back again, blinks the left eye, and bobs his head up and down, just once. He turns his head horizontally, then vertically, blinks his right eye, closes the left eye, then both eyes, or leaves both half open. And so it goes for at least an hour. Only one thing is constant—his left foot stays up.

Bubo often perches on only one foot rather than on both. Is he trying to keep one foot warm by holding it close to his belly, thereby expending less energy by shivering to produce heat? If it is important for Bubo to have warm feet, it would also help to explain why he has evolved feathers on his feet and toes.

THE barred and saw-whet owls that live in these woods also have thickly feathered toes, but barred owls in Florida have bare toes. Leon and E. H. Kelso (1936), who examined feathering on feet of American owls, found that owls with short-feathered toes are found in a slightly higher percentage in cooler environments than in warm, humid environments. The densely feathered toes are associated with owls of the colder zones. Willow and rock ptarmigan of the high Arctic have thickly feathered toes. But why do northern ravens, chickadees, and ruffed grouse not also have feathered toes? (In the winter, ruffed grouse grow a fringe of tissue flaps along the sides of their toes, and these fringes act as effective snow shoes.)

Bubo's radically different postures at different temperatures support the idea that his feet are covered for thermoregulation. When it is cold he is always fluffed out, and he usually stands on one foot. When it is warm (above 29°C) he looks like a different bird altogether: he always stands on both feet, lets his wings droop, exposes his underwings, and sleeks down his feathers, exposing the full length of his legs. But what actually is his foot temperature? My attempts to stick a thermocouple probe under the feathers of his toes did not meet with resounding success. Perhaps a braver person with a more cooperative owl will someday find out.

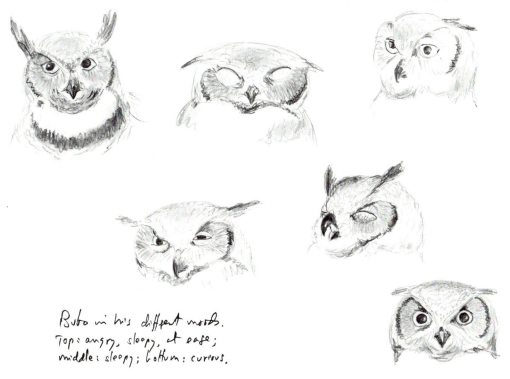

Bubo in his different moods.
Top: angry, sleepy, at ease;
middle: sleepy; bottom: curious.

AUGUST 24

The summer is over, and it is time now to face our responsibilities to Bubo. If he had learned to hunt, we could just leave him here. He might even still be near Kaflunk next year when we come back.

RECOVERY records of banded great horned owls on file with the U.S. Fish and Wildlife Service show that these owls usually remain near their hatching and breeding places throughout the year (Stewart, 1969). Two other banding studies of great horned owls also confirm that territories are held year-round, but if dispersal is necessary because of a dwindling food supply, then the burden of dispersal falls on the young (Adamcik and Keith, 1978; Houston, 1978). In Alberta, during years of snowshoe hare decline, the owls switch their diet to ruffed grouse rather than move away (McInvaille and Keith, 1974); they also raise fewer young or do not breed at all (Adamcik, Todd, and Keith, 1978). Ultimately, if prey switching and reproductive strategy still do not result in a balanced energy budget and the owls move away, then they may move a long way. A dispersal of up to 837 miles from the natal nest has been recorded (Adamcik and Keith, 1978).

There is about a 50-50 chance, though, that the young will not survive the first year. They suffer the highest mortality. In one study, first-year mortality of great horned owls in the wild varied from 46 percent (Stewart, 1969) to 55 percent (Adamcik and Keith, 1978), with the oldest wild birds living for thirteen to fourteen years. These data are based on several thousand banded birds and nearly eight hundred recoveries. Similar results have been obtained from a population of tawny owls banded in Belgium, where the individuals also stayed in the same territory, although they did not breed every year. One banded individual lived to nineteen years of age (Delmee, Dachy, and Simon, 1981).

THE ecological arguments for releasing Bubo into the wild are weak, because great horned owl populations are limited by food resources and habitat. Having found a piece of real estate with a plentiful food supply, an owl may have to fight a prior occupant for the right to keep it.

Annie Dillard has said: "Cock robin may die the most gruesome of slow deaths, and nature is no less pleased, the sun comes up, the creek rolls on, the survivors still sing. I cannot feel that way about your death, nor you about mine, nor either of us about the robin's—or even the barnacles'. We value the individual supremely and nature values him not a whit."

In the final analysis, Bubo is no longer just any owl to me. He reminds me of the fox that appeared to the little prince in Antoine de Saint Exupéry's *The Little Prince*. The little prince had said the fox was pretty to look at, but the fox replied he still could not be played with because he has not yet been tamed. To be tamed, the fox said, means to establish ties. If the prince would tame him, then the fox would no longer be like a hundred thousand other foxes. If he would tame him, they would need each other, and each would be to the other unique in the world. Bubo had been tamed by me and I had been tamed by him, and now he is unique to me as I am to him. He is no longer a statistical owl. Whether or not he went back to the wild would make no ecological difference. His not being there only meant that some other owl who would otherwise have died or not reproduced would take his place. My main concern, then, is how will this particular owl, Bubo, be best taken care of.

Given the various options, we decide to take Bubo to a raptor center specializing in rehabilitating injured birds. There he might be taught to hunt and eventually be returned to the wild with a fair chance of establishing his own territory.

Having made the decision, I show Bubo the piece of squirrel I had put into a large cardboard box lined with fir boughs. Unsuspecting, he immediately hops right in and I close the covers [a "mistake" he never again repeated in his life]. I expect him to tear the box to shreds, but he is as docile as a drowsy cat as I jostle the big box through the woods down to the road, to begin the journey.

The lady in charge of the predatory birds loses no time in letting me know my error, after I told her I took him from the wild. I am told in no uncertain terms that I have been "messing" with nature. Indeed, I have, and seeing her side of the argument I feel properly chastised. And so I remain silent.

She puts on a pair of thick, heavy welding gloves with sleeves up to her elbows, and then she rips open the cardboard box none too delicately. Bubo looks up meekly. I look at Bubo, then at Margaret. There are tears streaming down Margaret's face. We would both miss our friendly owl. The big gloved hand reaches in and grabs Bubo by the legs. Swinging upside down with futilely beating wings, Bubo clacks his bill furiously and turns his head, looking around helplessly with his bulging eyes. As he is dangling like a chicken to be slaughtered, she says: "There, maybe soon enough we'll get you to hate people." I suspect the words were more for our benefit than his. She did not personally know this owl. She could be objective, and should be.

She tosses him into a roomy wire cage and slams the iron door shut behind him. Other cages near him hold a red-tailed hawk recovering from shotgun wounds, numerous other great horned owls, a Cooper's hawk missing one wing, and a very sad-looking raven. I hope these rescue attempts are not construed as conservation, which is served most effectively by catering to statistical, unseen animals. It is here where our conscience and efforts should focus.

Bubo's second lesson here would be to get him to try to catch live mice released into the cage. Becoming proficient at catching mice, he

would be offered rats, and then possibly graduate to rabbits. That is
the plan. I am told that I would receive a call in two weeks telling me
whether or not he could be rehabilitated.

SEPTEMBER 16

The expected call comes today. I have been anxiously awaiting it. The
news is not good. The verdict is that Bubo is not earning his black belt
in prey catching. He does not even attempt to catch meek and slow
laboratory mice released into his cage. I find this news very strange
and puzzling. He had never balked for an instant at Kaflunk when a
mouse or some other animal was catchable. Nevertheless, Bubo is
pronounced "incorrigible." I am told that he cannot be released, because if hungry he will search out humans and unsuspectingly approach them for the handouts he has been used to. Few people will tolerate a beggar owl. They will mistrust his intentions and kill him.
Perhaps. But he had never approached any human but me. Was he now
really destined to spend the rest of his days in the cage between the
one-winged Cooper's hawk and the sad raven? This seemed like a
waste of time and resources, and of life.

   I could now only hope to get him back some day and return him to
the wilds near Camp Kaflunk so he could take his own risks.

Reestablishing Friendships

BUBO passed the fall and winter in his cage at the Raptor Center at the Vermont Institute of Natural Science in Woodstock. He refused to approach anyone there, even those holding out food. Six dead mice were left for him on a shelf every night. He waited until dark to retrieve them, and to eat them alone.

He now has a leather strap dangling from each leg. These jesses had been put on to hold him forcibly on the fist so that he might serve as a demonstration owl to the public. I know Bubo would have resisted fiercely—there must have been some heroic struggles! Surely he hated all attempts at "rehabilitation." Nevertheless, everyone urged me to keep him incarcerated at the Center because he could "never make it alone." But my rationality has its limitations and is often replaced by rationalizations, as in this case. I wanted Bubo back, because if indeed he could *not* be rehabilitated at the Center, and he therefore *had* to be caged or held fast by jesses, then I might as well be the one to do it. I do not like to see animals institutionalized.

When I persisted in trying to get him back I received one further warning: he is now an adult. "You will find him changed. He will be unfriendly, indifferent, and aloof. Owls become that way when they grow up." Whatever he had become, I could not imagine that Bubo had changed *that* much. In any case, I wanted to resume my mobbing study, and it was time to introduce him to various birds in the woods near Kaflunk. For the sake of science, I was ultimately granted permission to take Bubo back.

I could not go to Kaflunk until June 1, but Bubo would be delivered to my office at the Zoology Department at the University of Vermont in Burlington on May 18. To prepare for his arrival, I built a 30-cubic-meter flight cage in my backyard at home. It was walled with barn boards on two sides and roofed over to create a dark shady nook, and it enclosed a large pine tree trunk with all its limbs. Two horizontal poles at opposite ends of the cage served as additional perches.

My plan was to look at the interactions of crows and the owl more directly. I wanted to know if young crows instinctively mob owls, or if they do so by learning on their own or from others. I decided I would have to get "inexperienced" chicks out of a nest and raise them myself.

Crows, however, are protected birds. According to the U.S. Fish and Wildlife Service's Form 50 CFR 10, red-winged blackbirds, brown-headed cowbirds, American crows, and common grackles are migratory birds "protected by the Migratory Bird Treaty Act (16 U.S.C. 703-711)." Form 50 CFR 21 (p. 3 of 11 pages, section 21.23) explains that "A scientific collecting permit is required before any person may take, transport, or possess migratory birds, their parts, nests, or eggs for scientific research or educational purposes." Furthermore, "unless specifically stated on the permit, a scientific collecting permit does not authorize the taking of live migratory birds from the wild." In short, taking a live crow or a blackbird is strictly *verboten*.

It is very tempting to take a wild animal as a pet on the spur of a moment, forgetting that this impulse must be accompanied by a commitment to give constant care to the animal for its lifetime. Everything that the animal needs that nature in her vast domain provides cannot be easily replaced. But crows are suitable pets on many accounts: They are in no danger of becoming rare. They do not need to be caged. They thrive on cat and puppy chow. They are affectionate, entertaining, and resourceful, and finding them in the wild, for free, is an educational experience. On the other hand, buying a rare parrot for $2,000 from a pet shop merely supports wildlife destruction for the sake of keeping an exotic ornament in confinement. I wish everyone could keep a crow instead. They would be infinitely more rewarded by the experience. But it is illegal.

Anyone, however, can *kill* a crow—or thousands of them, for that matter—for the price of a hunting license. Referring to my 1985 Vermont Digest of Fish and Wildlife Laws, I learned that "crow" is classified as a game bird, along with ruffed grouse, pheasant, and quail. For an eight-dollar licence purchased at the nearest town office, anyone over the age of sixteen (or under, with a parent's permission) can shoot, per day, four ruffed grouse, two pheasants, and four quails, from the last Saturday in September until the end of December. As to crows, which are "protected" by the aforementioned Migratory Bird

Treaty Act, there is "no limit." Hunters are allowed to shoot as many as they want not only every day of the regular hunting season in the fall, but also in the spring between March 14 and April 30 (which is near the time that they hatch their young); at that time, of course, there is risk of killing the young through starvation by shooting the parents. Between August 16 and October 29, during the fall hunting season, crows aggregate in communal roosts and are again easy targets. The state of Maine also has "no limit on crows," and the spring crow season is the same as in Vermont, but the fall season is even longer—from July 16 through September 29. What is the "rationale" for these specific open seasons? Who eats crows, anyway?

I got my state and federal permits to study two live crows. But it required the processing of reams of paper, months of waiting, and the purchase of a state hunting license that served as a fee.

MAY 19–23

My two baby crows have their own room in our house, safe from the cat. They sleep in their nest in a cardboard box. And they wake every hour or so when I feed them. Bubo is in his new cage by the garden. He perches immobile in the darkest corner, and his huge pale-yellow eyes are always open.

His massive curved bill is partially hidden by a bushy mustache of bristly white feathers whose wiry ends look as though they have been dipped in India ink. Directly under his bill the feathers are almost horizontal, so his throat looks as if he is wearing a ruffled collar. And just beneath this collar there is now a large white bib surrounded by accentuating black-tipped feathers. The smooth surface of his breast is a mesh of long, broad, overlapping and interlocking feathers with wavy horizontal bars on a white-to-creamy background. The white is nearly transparent; you see the outlines of the dark bars of the feathers being overlapped through the white above them. From up close this gives the breast almost a shimmering appearance, like a Jack Turner watercolor with many thin layers of pigment. The bars blend into each other near the top because they are so close together. But lower on the chest and abdomen they become increasingly farther apart and more individually distinct.

His 0.6-centimeter-thick, steel blue-gray talons are partially hidden by his flowing breast feathers. But their length, now 3.8 centimeters, precludes them from being hidden entirely. He looks like a beautiful owl, all right—a stuffed owl. He acts lifeless. Is this really Bubo?

Although his feather coloration is like that of the Bubo I knew, it could also be like that of other great horned owls from New England. There is one mark, though, that might positively identify him. The previous summer [see May 19] Bubo had flown against a wire of his aviary at Kaflunk, leaving a scar at the base of his bill—a small diagonal indentation. Checking for this mark I now find it again, but it is one centimeter farther down toward the tip of his bill. Apparently the bill grows at the top and wears down at the front, like a beaver's teeth. There is no doubt, then, that this is really Bubo, at least physically.

He does not recognize me when I enter the cage to talk with him, or else he no longer cares. Instead of making the little grunts of contentment in my presence as he always used to, he now makes a faint dull hiss. When I walk close to him, he partially opens his bill, expressing displeasure or fright, and he chirrups irritably when I put my gloved hand near his feet to try to lift him onto my fist. I persist in trying to get him onto my hand, and he sleeks back his feathers, raises himself erect, and flies to the perch farthest away from me. He is irritated by me, and maybe bored as well.

While I am standing in front of him trying my utmost to reestablish a meaningful relationship, he gazes past me into the distance, or showing me the back of his head for amazingly long periods of time. I am disappointed that I am of no more account to him. I had hoped we would still be good friends. But I had been forewarned. He is indeed a very changed owl.

For four days every morning and every evening, I have been tempting him for an hour or so with pieces of squirrel and muskrat meat, with fur and without. Bubo pays no attention to these delectables. He does not even take the best I have to offer when I hold it directly under his bill. He chitters irritably instead. I do not even get the satisfaction of seeing him eat.

MAY 24

Today I have something new, and hopefully interesting, to offer Bubo—two young mice, live ones. His cage is not mouseproof, and so I secure one of the rodents by a thread tied to its tail. I do not expect him to catch the critter, because I have, of course, been previously informed that Bubo refuses to try to capture mice, and could not be taught to do so. After all, he is incorrigible.

The mouse sniffles about on the dirt floor and shuffles around, clumsily trailing the thread. Bubo looks down from his perch, slowly cocking his head this way and that. After about a minute, though, he loses interest. He lifts his head and pulls it back down into his shoulders. He has seen enough. His gaze shifts blankly into the distance. Six days without any food—and he still shows no interst in *a live mouse*! He indeed acts as I have been told, and I do not understand. But I do not give up on Bubo so easily. One cannot, if one is bound by affection.

Maybe he should get a closer look at what he is missing. I set the mouse opposite him on his horizontal perch. This mouse, which has only known a lab cage, is not very acrobatic, but it manages to hang on. Indeed, slowly and cautiously it "walks the plank" forward, looking down now and then to the abyss below, but not to the real hazard ahead. It is concerned with the present crisis—and precisely because it is so studiously trying to avoid it, it courts disaster ahead.

Bubo, meanwhile, cocks his head again, showing casual interest. When the mouse is a few centimeters from his toes, Bubo shifts his weight slightly from one foot to the next, as if uneasy and not knowing quite what to do next if the mouse should come any closer. It does. When it is about to reach a less exposed, more shady place (directly under Bubo's bulk and feathery toes), Bubo lifts his nearest foot and neatly plucks the critter from the limb. The tiny mouse squirms briefly as the talons curl securely around it. Bubo then looks up and once around, jerks up foot with mouse, closes his eyes, and in one bite crushes its skull. The mouse hangs limply in his bill for a few seconds, and then Bubo drops it to the ground.

He glances down only briefly, and then goes back into his trance. Eventually he does swallow small pieces of the mouse that I tear off and hand to him, though he is somewhat phlegmatic in eating his meal. This is not in the style of Bubo the glutton, the vanquisher of moss hummocks, the attacker of falling squirrels. Something must have happend to him in captivity. Instead of having been rehabilitated, he has regressed, precipitously.

Later in the evening he spontaneously hops to the other perch. That is something. The first hop is always the hardest. I am now hopeful.

I reenter the cage with mouse number 2 which is also trailing a thread from its tail. Bubo looks down at it. The unsuspecting mouse is twitching its whiskers, snuffling in the dirt in front of it. But as before, the danger does not lie on the floor in this new alien environment, because mouse number 2 is being watched from above, by Bubo. After twenty seconds of studious contemplation, he has made a decision, and acts on it. Dropping from his perch, he hits the mouse with both feet. Hurray!—I am elated. After a brief look around, he makes an effortless leap back up to his perch carrying the mouse in his talons. In one gulp he swallows it head first. The mouse's tail briefly hangs out the right corner of Bubo's mouth, and slowly slides in along the corner of his bill. I cut the thread and give a silent cheer. This owl *will* come back to life again, I am sure.

May 26

Throughout the day Bubo sits stolidly on his perch, as usual. Only his eyes show signs of life. They are constantly open. He turns to look in the direction of distant barking dogs, cars, kids. He watches gulls fly in the distance, and his pupils contract to small black points as he follows their flight in the sky. Spending one or two hours beside him every day I notice his eyes and his face a lot.

His eyes, set in a 12.5-centimeter-wide head, are large. They are 1.9 centimeters wide on the exterior alone, larger than my own, and the eyeballs bulge well beyond the surface of the head. His upper lids have short bristly "eyelashes," and the pale yellow of his eyes is set in broad russet facial discs framed in a ring of black. The rest of the head is mottled in chocolate brown, and topped off with nearly a dozen sienna brown "ear" tuft feathers tipped with black.

At 7:30 P.M. he hops to a branch near the front of the cage. I extend a dandelion on a long stem through the wire to him. He reaches over, chitters in a whisper, and pulls out some petals while shaking his head in a gesture of distaste. Still, he persists and pulls out and scatters a few more petals. Is he playing? Hungry? I offer him some squirrel meat. He holds it limply in his bill, then drops it. I offer it again. This time he flies to the ground with it and walks to a corner of the cage, where he hides it in the grass. He is still not hungry, but life is returning to him.

MAY 28

For ten days now I have talked to Bubo for at least an hour every evening and again every morning at dawn, the times when he is most alert. Perhaps my attention to him is paying off because now, instead of opening his bill and hissing when I come close, he bobs his head up and down and makes little grunting sounds. I can hold my finger near his bill and he closes his eyes and caresses it with gentle nibbles. When I offer him my naked forearm, he nibbles it as well.

I slowly let my finger wander to the top of his head to reciprocate the caress, but he chitters in irritation, sleeks his feathers, and delivers firm (but by no means crushing) bites to my fingers. I get the message, and do not persist in my attempts at familiarity with him. It is not yet time. As Merlin said to the Wart who wanted to make friends with the tawny owl, Archimedes: "Let him alone . . . perhaps he does not want to be friends with you until he knows what you are like. With owls, it is never easy-come and easy-go." Especially with Bubo.

Bubo playfully nibbling on author's fingers.

Return to the Home Territory

JUNE 1

I am bathed by the warm light of the kerosene lamp and I hear the wind in the trees outside. No cars. No airplanes. From the distance, down off the ridge, I faintly hear frogs. What peace. It's good to be back in Kaflunk.

Everything in the cabin is almost exactly as it was when we left it at the end of last summer. The tools, the .22 rifle, the utensils, the food, the ammunition, the clothes, the bedding—all are in their place. Only, the mice have chewed up the soap, and the beer can that held down a small note on the table saying "Make yourself at home, but please leave it as you found it" is now empty. Now there is an additional note, saying "Thanks. We loved it here. We appreciate your hospitality," and it was signed by many people I did not know, who left me their phone numbers and addresses. It is so every year. Backwoods hospitality is a tradition already observed by Thoreau in his 1846 trek to Ktaadn: ". . . every log hut in this woods is a public house."

Margaret is visiting her parents, showing off our handsome new six-month-old son, Stuart. She will come in a few weeks, which happens to be after the peak of the blackfly season. Thoreau also had a few things to say about these flies, *Simulium molestum*. He said they "make travelling in the woods almost impossible" and they are "more formidable than wolves."

It was still daylight when I came up. The wild apple trees along the trail were laden with pink blossoms. A ruby-throated hummingbird was hovering, darting from one flower to the next, feeding on nectar. Cedar waxwings were eating mouthfuls of the petals, making high-pitched squeeking noises. I had seen them eat fruit in the fall, but never before had I seen them eat flower petals.

The leather-leaf, high-bush blueberries, and bog rhododendron I planted last year are growing. The rhododendron is already aflame with purple bloom and attracting many bumblebees. The logs of the new cabin have weathered to a soft gray over the winter, blending in with the muted greens and browns of the pines, spruces, and maples. The hills show patches of fresh pea-green and yellow as different species of hardwoods are leafing out, and the blue violets on the forest floor are in full bloom.

Walking up the trail I heard oven-birds, white-throated sparrows, and Nashville and magnolia warblers. These birds are now starting to lay their eggs. The hairy woodpecker has small (but noisy) young, and the young ravens had already left their nests two weeks ago. I see almost no insects except small, light-brown moths that are occasionally flushed from the wet leaves near my feet. Thoreau said that the forest "is even more grim and wild than you had anticipated, a damp and intricate wilderness, in the spring everywhere wet and miry." To me it seems neither grim, wild, nor miry. It simply seems warm and hospitable, the blackflies notwithstanding.

Bubo seems no worse from his ride from Vermont. He stopped struggling after I finally cornered him in his cage and captured him in a (now-dreaded) large cardboard box. He is now in his old aviary next to the window where he stayed last year before he could fly. He came up to the window and craned his neck to look into the cabin, chuckling and bowing to me, plainly showing excitement and maybe even pleasure. He remembers! Whenever Bunny comes close to the window, Bubo ruffles his feathers, hisses, and clacks his bill. He even has the same distrust of Bunny as before.

Aside from Bubo and Bunny, I brought along the two young crows that are now feathered-out and almost ready to fly, and who ravenously eat canned cat foot and other delectables that I chew up and moisten for them. They are now three meters up in the mountain ash tree in front of my window, in an artificial crow's nest safe from the cat and the owl. The eventual but inevitable crow-owl-cat interactions should tell me something about the unlearned versus learned aspects of mobbing behavior. But I can release Bubo and Bunny in the crow's presence only when they are nearly adult and able to fly well. In the meantime Bubo and Bunny will have to share the cabin with me, as the crows continue to mature outside.

Just as the touching of electrodes on the brain can retrieve memories of long-gone events and feelings, the crows' musty smell gives me

The crows Theo and Thor at Kaflunk.

an instant trip back to the childhood magic when I had my first pet crow. The feelings I had then and that I could feel with that strong force only during childhood nevertheless often still echo themselves in fleeting flashbacks. Usually when I experience these intensely pleasant moments it is not because of new things I see, but because of something old retrieved. Perhaps I am searching in nature not only for the adult intellectual challenge, but for the stimulus that taps these feelings still hidden within the subconscious child. I feel sure that my deep interest in nature was stimulated by my very early intimate contact with wild things, such as my pet crow, Jacob, who was my first close connection to *wild* nature.

JUNE 2

I have been busy this morning unpacking and getting the cabin back in order. Bubo came in through the opened window from his aviary to join me. He is now asleep, as in the old days, on a 2×8 beam. His lower lids are drawn up over his eyes, but he slowly opens his left eye when I talk to him, while his right one remains closed. I am reminded of the 24-hour ultra-runner who tried to sleep and run at the same time, by wearing an eye patch. Maybe Bubo is only *half* awake. His ear tufts are erect, and his breast feathers are fluffed out. Two toes stick out horizontally from under his belly feathers; one foot is retracted up into the feathers. His talons do not grip the beam but stick out in front, and he rocks almost inperceptibly from side to side. I feel good knowing he is up there, totally at ease in my presence.

Early this morning, when I got up, he was perched next to the windowpane, between the cabin and the aviary, peeking into the cabin. Finally, thinking, "What the heck," I opened the window. Bunny was sleeping on my bed, next to the window, but he did not stay that way for long when Bubo came onto the windowsill and started peering at him. Bunny looked over his shoulder and found himself in a staredown. Bubo was patient. After several minutes Bunny got up, meowed, and climbed the ladder to the loft to search for more private sleeping accommodations.

Bubo charged in, and with a graceful swoop he landed on the rafters where he is now. But first he spent about an hour exploring the cabin, getting reacquainted with his old haunts. He flew from rafter to rafter, floor to rafters, to the sink, woodpile, table, chairs, desk. The only place he avoided was the stove (which was still quite cold); perhaps he remembers his lesson of a year ago. He shredded a roll of toilet paper, attacked the leather gloves, tore a washcloth into strips, and carried a T-shirt up to the rafters for future savaging. He paid no attention to me all the time that I was unpacking, sweeping the floor, and tidying up.

Having returned to the premises and having explored and ransacked them, he eases into contentment on his old perch near the door, alternately preening and napping. His comfort—perhaps his happiness—at finally having returned to his home territory is palpable. He is clearly alive again, and from now on, I am confident, he will become his old self, and much more.

When he preens he is fluffed out, and as he delicately draws a feather through his bill from base to tip, he closes his eyes. He reaches over his back, holds out a wing, and works the wing feathers on the inside of the wing. Or he may bend his head down sharply and do the feathers on his throat, then reach back to do his tail or shoulder, or he may lift a foot and work on the tiny feathers on his toes and nibble on his talons. Next, he pulls on a long breast feather, then shifts weight to the right foot, raises the left, and scratches the top of his head with an extended talon, working it back and forth the way a dog scratches behind its ears with a hind foot. As a break in this routine, which by no means progresses in any systematic sequence, he may stand still

Bubo on the rafters—sleeping, relaxing, preening, shaking.

for a few seconds while all of the body feathers slowly rise until they are nearly at right angles to his body. The feather raising is invariably followed by a shake. A dog usually shakes its head first, then its body, and lastly its tail. Bubo does the reverse. He shakes his tail first and his head last.

He reminds me of someone trying to shake out a coat without taking it off. Like a coat, his feathers retain a thick layer of insulating air. The coat hangs over his toes and flares to the sides and scallops around the tops of the wings that are neatly tucked in at the sides of the body, where they, too, act to hold in body heat. Unlike any of my coats, the insulation of his is adjustable to allow for various temperatures. When he expands his chest width from 15 to 20 centimeters, he is merely elevating his feathers to increase the insulating thickness. His breast feathers may also be slightly parted at the bottom, like a tuxedo, and by raising the feathers he separates the coat at the front, as if opening it from the bottom up. When thus "unbuttoned," he reveals a creamy white, fuzzy undergarment.

After getting things settled in the cabin, I go into the woods. It looks as though it will rain almost any minute. The gray skies highlight the fresh pea-green foliage now bursting out from the beech trees on the hills. Many tree species are now starting to leaf out. White wild-strawberry blossoms peek through the fresh grass sprouting near the cabin door. During the day they are visited and pollinated by small, unobtrusive bees of many species whose names I do not know but wish I did. The bees are as inconspicuous as the relationship they have with the birds and other animals that will later eat the berries that are the result of pollination. There is a huge web of unseen symbioses, mutualisms, cooperation, competition, and randomness. These themes are not necessarily mutually exclusive. To nature it is all the same, and only cause and effect remain constant.

We live in a remorselessly logical universe where causes have effects, where actions have consequences, and where one pays dearly for one's mistakes. There is no magic. Only mystery. We wish it could be a kinder, more forgiving environment where the innocent are spared and where horrors are prevented by some all-powerful God. But it *can't* be so. In an editorial in the *Boston Globe* (October 12, 1985), Gwynne Dyer points out why:

Imagine for a moment a universe in which tragedies didn't happen. When the engines of a jet airliner fail on takeoff, it does not crash at the end of the runway and burn 150 people to death. Instead, it just wafts gently to the ground, because God loved the passengers and chose to save them. But if that were all that happened when aircraft engines failed, there would be no need for aircraft maintenance. Indeed, there would be no need for engines, or even wings—and people could safely step off the edge of cliffs and walk on air. The law of gravity would be suspended whenever it endangered human lives. So would all other laws of nature. . . . In such a universe there could be no science or technology, because there would be no fixed natural laws on which to base them. . . . There could not even be logic, since the same causes would not invariably have the same effects.

And it is, of course, precisely this cold-blooded logic, this unforgiving aspect of the universe that has created the beauty that we see around us. It has created beautiful spring flowers, gorgeous butterflies, owls, and human intelligence. It can also create as close a heaven on earth as the universe will allow, but only through scientific literacy. As Daniel E. Koshland, Jr., said in another editorial, in *Science* (October 25, 1985): "Scientists will be denounced for trying to introduce cold-blooded reason into an area in which warm-blooded humanity is supposed to reign supreme. But warm emotion frequently gives way to hot-headed anger and even bigotry. The scientific method has been the most effective means of overcoming poverty, starvation, and disease." "The" scientific method, whatever it may be to different people, has one constant: hands-on contact with nature to determine from a quantitative perspective how nature works.

WHEN I return at 1:30 P.M. Bubo is still on the same beam where I had left him at 10:30 A.M. In a few moments, though, he stretches. He becomes active again now near midday, apparently because I have returned, but he does not beg for food. After giving the whole room several inspections without leaving his perch, he hops onto the cast-iron sink and retrieves the green washcloth. He carries it all about the cabin for twenty minutes, occasionally stopping to render it to shreds. He pounces on a dead mouse I drag across the floor on a string, and after swallowing it he returns to the same beam.

His lower lids are raised again. I, too, am relaxing, reading at the table just below him. Is he really asleep? I call up in a quiet voice: "Buuu-bo." His left eye opens slowly. He answers in two or three soft, barely audible chuckles that make his shoulders jiggle. I call again. He reacts the same way, until the sixth time. Then he comes to rest, except for his digestive system.

A dull-sounding "plop" interrupts my concentration. Bubo has regurgitated a pellet. It is bright green. Can it be? Yes; it is composed of moist compacted washcloth. It was produced in only one hour. This must be a record time. There is much literature on digestion in owls, and according to all that wisdom I should not have seen that pellet with the green marker for another nine to twenty-four hours.

The owl's diet consists of easily digestible meat that must be separated from the indigestible bones, fur, and feathers. Food that is digested in the stomach must pass on to the intestines to be absorbed, while undigested wastes must pass *up* to be discarded. The two-way traffic in the digestive tract creates a potential problem if meals are frequent, because a second meal might interfere with a prior meal that has not yet been disposed of. Indeed, a great horned owl given many mice can process them faster if it gets them all in one portion than if the same meal is divided into two or three equal portions (Fuller and Duke, 1979). Owls do not snack. They gorge. That suits their physiology. But dish rags?

How does an owl's stomach distinguish a piece of washcloth that must be routed up from a piece of squirrel hide that it keeps down for further processing? How is the pellet-formation assembly line stopped and rerouted? What internal sensors regulate the traffic and decide on what travels where? I don't have answers. Perhaps somebody will take the question on as a puzzle.

In the afternoon Bubo has another session of activity. As before, he becomes active when I do, and when I am working at my desk he returns to his perch above me and preens, and snoozes, and. . . .

Bubo awakens and stands up expectantly, and then I hear a holler from the trail below the cabin. Jim Marden, my graduate student, and his wife Paula are coming for a visit. They come in and are almost at ease when, from just above them on the timbers of the cabin, resounds the full-bodied, resonant hoot of the great horned owl. Bubo has finally found his voice! He can still hoot, after all—and *how!* He stares malevolently at Jim (in Burlington he had ignored Jim completely).

More hoots follow. And more. With each hoot his white throat bib bulges out as if it is being inflated. His head feathers are sleeked back, and his "horns" are erect, as is his tail.

It is an interesting and novel experience to have a great horned owl hoot next to your ear, but after the fiftieth time the novelty wears thin. This is not the soft, friendly hoot he used last summer when we met by surprise in the woods. He had looked friendly and at ease then, nor had he hooted malevolently at strangers at Kaflunk. He had only been upset by them. Now his hoots express power, determination, and anger. He focuses on Jim, ignoring Paula. But his anger is general, and he attacks my glove, and then even my legs. The Mardens had come to stay the night, but after Bubo shows no signs of calming down in an hour or so, they become uneasy and leave to spend the night elsewhere.

As they walk back down the trail, Bubo hops down from the rafters in order to watch them better from the window. He continues to hoot until both are well out of sight, and a half hour later he shakes himself, preens, and is silent and at ease once more. I think Bubo has just successfully defended what he now claims as his home territory.

Life in the Cabin with Bubo

JUNE 3

It already seems late in the day. The robin started singing about two hours ago while it was still dark, and it has long since stopped. The robin was followed by the Swainson's and hermit thrushes, and they, too, have already stopped singing. The white-throated sparrows also started early, but they still continue their serenade. The ovenbirds occasionally interrupt the melodious singing of the other birds with their loud, raucous stanzas. Blue jays on their morning errands scream to each other at odd intervals in the forest, and the Nashville warbler near the blueberry patch by the cabin incessantly repeats his monotonous cascading call. The late risers, the rose-breasted grosbeaks and the scarlet tanagers, are only now starting to sing. It is 5:50 A.M.

Before Bubo came into my life I usually started my day a little later, but thanks to his rearranging my schedule, I have already had breakfast, and I am enjoying my second cup of coffee in front of a warm fire in the wood stove. Up here on the hill in Maine the mornings are still cold.

All was quiet in Bubo's cage during the night. But at the first flush of dawn he tapped on the windowpane next to my head. He was persistent, and when I opened the window Bubo hopped onto the sill. He studiously surveyed the room for a few minutes. The coast was clear. Bunny was fast asleep on the bed, having retired from the nocturnal hunt. Bubo launched himself across the cabin, and landed on one of the beams.

Many large birds have difficulty lifting off in flight. Not Bubo. He has long, strong legs hidden under his flowing garb, and he can leap-fly straight up from the floor and land on the beam 2.3 meters above his head. It looks effortless. It also looks incongruous in one who waddles like a goose and who reminds you of a short, stubby, and overweight old man.

7 A.M. Bubo continues to be active, even though the sun is now shining brightly. He finds a roll of fish line and works that into a frazzle. He hops from floor to beams and back down. He preens and shakes, explores some more, and then spends much time peering out the windows.

And then again, as if on second thought, he puts his head down, spreads his wings, and hisses at Bunny who is reclining on the bed. Strangely he does not clack his bill, and he even *advances* toward the cat with his head close to the floor and his body rocking from side to side. Bunny, apparently remembering Bubo from last year and knowing this is all bluff, does not budge. Bubo then nonchalantly waddles about elsewhere on the floor in a slow trot, each footstep making clicking noises as his talons hit the pine boards like the studs on hobnailed boots.

Yesterday his main focus was the inside of the cabin. Today he is almost finished with cabin inspection and spends much more time looking out the windows. Perhaps I need to arrange some entertainment for him inside. I have two live crayfish that might interest him.

Crayfish number 1 is lying on its back on the floor, moving its legs slowly and flexing its abdomen. Bubo stares at it from the beam above. His stares alternate with occasional violent shakes of the head, as if he is saying in irritation "Yuck! What a crawly, unappetizing thing." But curiosity gets the best of him. After staring for five minutes, he hops down for a closer look, still giving his head an occasional quick shake as he walks closer. By the time he has made a close inspection the crayfish has stopped moving its spindly legs. Bubo loses interest.

I now put the crayfish into several centimeters of water in the dishpan. The water revives it and Bubo's interest is renewed. He walks around and around the dishpan, weakly reaching in for the crayfish with his bill, but coming up with water instead. He seems to like the water, takes a drink, and hops in, making the high-pitched squealing noices he used to make last year whenever he took a bath.

The dishpan is too small to splash in, and he rediscovers the crayfish at his feet. He grabs it with his bill, crunches it, hops out, and crunches it some more. Mmm—not bad. He swallows it. Nineteen and a half minutes have elapsed since he first saw the crayfish.

Half an hour later I present crayfish number 2. This time he goes through the same maneuvers, only this crustacean is committed to the stomach almost exactly three times faster.

When I return to the cabin later in the morning after a four-hour absence, Bubo is fluffed out and asleep on the rafters, and Bunny is also asleep, on the bed, as usual. They have the utmost respect for each other, and they have made their peace out of mutual fear.

Bubo has not been altogether unoccupied while I was out. My second green washcloth is on my desk, torn into long shreds. I look up: "Bubo, did you do this?" Bubo looks at me, and as if on cue stretches his neck, opens his mouth wide, and out pops a large green pellet composed of a 30-centimeter-long crumpled strip of green cotton terrycloth enclosing exoskeleton fragments of crayfish.

JUNE 4

Last night Bubo went into his cage on his own. Once inside he perched calmly for a half hour, facing the woods where the hermit thrush was singing. I went to him with a mouse in my gloved hand. What a thrill—he flew to my fist! I put him back on his perch, and he came again. I got a few bits of muskrat meat, and he came to me as he had always done before. He stayed on the glove without protest.

Bubo is getting more friendly (or habituated) by the day. When I come back to the cabin he now always comes down from his sleeping perch to play with paper, to pounce on shoes lying about, or to tear any shirts he can find. When I read he comes next to me to perch on the table so he can nibble at the pages and on my fingers. I use the opportunity to scratch his head.

7 P.M. Like last night, Bubo flies back out into his enclosure. An hour later I still do not hear any sound of him moving from perch to perch. Peeking in, I see him fully awake. He faces in one direction for a few minutes, turns his head quickly, and then holds it completely still again. His physical restraint seems to contrast with his apparent inner tenseness.

8:20 P.M. The light is beginning to fade. The bell-like melodious piping of the hermit thrush comes from the evergreen forest at the edge of our clearing. Several yellow-rumped warblers are trilling from near the tops of the spruces. An ovenbird calls once or twice in its raucous voice, making the yellow-rumped warblers sound weak and dainty in comparison. A nighthawk sails over, very unusual for this area. And in the blueberry patch, a bumblebee is humming, collecting its last load of nectar for the day, and unwittingly pollinating hundreds of blossoms.

8:50 P.M. Darkness is falling. The birds have stopped singing. Bubo's vigil has ended, and he is now flying from perch to perch.

JUNE 5

Bubo wakes me a little before 4 A.M, when he starts pouncing from perch to perch, rattling his cage. The sky is already lightening in the east. A half hour later, at exactly 4:34 A.M., within one minute of yesterday's time, Bubo taps on the windowpane, asking to be let in.

Bubo, at the open window, surveys the room for several minutes. Before he flies off in any direction he considers a number of alternatives. He never jumps without lengthy deliberation. This is part of his appeal.

Through half-closed eyes in the murky light I see Bubo lift my long-sleeved sweatshirt off the floor, trailing it in his talons as he flies to the rafters. I might have gone back to sleep, but the sound of ripping cloth jars my drowsiness. I hear the beating of his pinions in the hollow confines of the cabin. Then there is the occasional clank of a dropped spoon or a tipped-over glass. My sleep becomes light at an unnaturally early hour. Bubo, on the other hand, is wide awake and misses few opportunities for mischief. Now he discovers my running shoes, but these are too tough to tear, and they are not very noisy when dropped. He leaves the shoes and tries the broom next. He has moderate success in dissecting that but bolts away when it smashes to the floor. As usual, his greatest achievement in over-all effect is with a roll of toilet paper.

Something crashes on the floor. Bubo's talons reverberate like spiked shoes as he runs away from it across the floor. Then all is quiet. He is on his perch up on the rafters, preening, until I get up and he again resumes his games, now according to my schedule.

I saved the tail of the woodchuck he fed on last night. It is a handy toy to drag along the floor by a thread. Bubo sees the opportunity for fun and runs after it at once. He runs faster and faster as I keep it just out of his reach. Finally he "catches" it in his bill, but I yank it away. He chitters loudly in irritation and resumes the pursuit. Round and round the cabin floor he goes, now making wing-assisted leaps. This is a bigger challenge requiring all of his tools, and he flies and pounces with his talons held forward, his head back—the classical raptor pounce. Success. I let him keep the tail, and he wastes no time swallowing it.

Bubo makes at least two sounds with his bill: the snapping used as a threat, and a soft smacking used to express satisfaction. After swallowing the woodchuck tail, Bubo bill-smacks. I do not know if this serves a function in communication. But I understand: the capture and eating of the tail was a satisfying experience for him.

JUNE 6

We are expecting a reporter today who had heard some rumors about someone living out in the woods with crows, and owls, and other things. I build a fire in the stove and put on some water for coffee. For once, the floor is swept. Shreds of toilet paper have been picked up. Droppings have been wiped off the floor, chairs, and table. Even the dirty dishes are neatly piled. In short, the cabin is as ship-shape and ready for company as it will ever be. Bubo, meanwhile, is still perched on the glove that I once used while forcibly holding him by his jesses (and how he objected!), and which he has been savaging since 5 A.M. But now, at 11 A.M., he is beginning to look drowsy.

He does not stay drowsy for long. I am just settling back and admiring my housework when I hear footsteps outside and a cheerful "Hello?" Bubo hears, too. Instantly he cranes his head down to about the level of his toes in order to look out the window from up on the rafters. The sight must make a deep impression—he drops the glove that has not left his talons since dawn.

The reporter, a man of about my age and build, is approaching the door. Bubo's body becomes horizontal, his tail is raised, and the feathers of his head sleek back, with ear tufts straight up. His eyes change from round to almond-shaped, and his white throat patch puffs out as he hoots his full-volume challenge, "Whoo who-who-who whoo-who whooo." There is a strong emphasis on the last syllable. That means he is upset. The reporter stops in his tracks, but I invite him inside. As he steps to the doorstep Bubo is already flying about the cabin, making a shambles and hooting all the while.

The reporter, a little hesitant now, sticks only his head through the door. Bubo rushes at him. That does it. As the interview proceeds outside, Bubo's calls continue to reverberate from inside the cabin, interspersed with occasional crashes. Out of the corner of my eye I see papers flying about and hear dishes clanging and boxes tumbling. The reporter diplomatically suggests we find a more quiet place, and one with fewer blackflies. I am now worried, but not for the reporter, who could ward off the owl by holding a stick in front of him. But what about my young son, Stuart, when he comes?

We retire to the clearing and climb into the partially erected log cabin. Here we are able to evade Bubo, but not the blackflies. They gather in swarms to slake their demonic thirst on any patch of skin they can find. They find quite a few, and leave purple, itching blisters.

Minutes after the blood-letting begins, the reporter remembers something he forgot in the car, and we descend the trail, slapping our faces, hands, and the backs of our necks. Very effective critters, those flies, in making the woods uninhabitable to most humans. I do not court these *Simulium molestum*, but at the same time I grudgingly give them their due in their role as conservationists. Few people like to settle on, and despoil, their territory. They help to keep Maine green. May they live forever. In the meantime, I slap quite a few of them.

Usually I need to hold a net and a note pad in one hand, and an instrument and a pencil in the other, while I patiently pursue bees or butterflies and while the blackflies, mosquitoes, deerflies, and no-see-ums doggedly pursue me through the day in relays. It was now the blackflies' turn, but my hands were free to defend myself this time. That was a treat.

A full three hours after the reporter has left, Bubo is still hooting incessantly in the cabin. I would never have guessed that the man would be so threatening. Maybe Bubo can be made to forget the intrusion by diverting his attention to the immediate. I call Bunny, luring him close to Bubo with some canned milk. As I had hoped, the cat banishes the stranger from Bubo's mind. Bubo sees cat, watches cat, and threatens cat. His concentration is broken. He stops hooting for a half hour, but after that the stranger comes back to haunt him, and I again have to endure the hooting. I try the cat remedy again, and the second dose finally cures him.

Most birds open their mouths to sing. But Bubo always has his bill

Bubo hooting while defending
his territory.

tightly closed when he hoots, and his throat puffs out like a frog's,
fully exposing a large, exquisite, white feather patch that shines in the
dark like a reflector.

BUBO prefaces most of his hoots to me with grunts, but hoots to strangers are never prefaced this way. John T. Emlen (1973) similarly noted
such grunting in a male in the wild who was communicating with a
female:

> On several occasions when the birds were close by, I could hear the
> male uttering a rapid irregular series of soft short grunts between
> hootings. These waxed and waned in intensity until the female
> hooted, whereupon he promptly responded with his hoot. Appar-
> ently, the male's persistent grunting stimulates the female to hoot,
> which, in turn, triggered his hoot. When the female did not hoot,
> the male continued his soft grunting for 10-15 seconds and then
> hooted independently. The grunting suggested increased agitation
> in the male, held in abeyance until released as a full hoot by the
> stimulus of the female's hoot.

Bubo does not only have many different kinds of calls, he also has different nuances in any one call. And these nuances are tinged with meaning. It has taken me a long time to notice the differences, and even now I understand only some of them. His standard "official" hoot is composed of seven notes. But occasionally he gives only six notes, or even five. When he is angry and hooting at a stranger, he always gives the seven-note hoot, but with a strong emphasis on the last note and an increase in pitch and volume. It is like a "Hello" that is fairly shouted, not as a greeting, but as a challenge. And when he issues his challenge he stares directly into the antagonist's eyes. On the other hand, when he gives a friendly "hello" hoot, there is a major drop in pitch and volume, and especially the last note is then soft and faint. When he is friendlier still, he may drop the last note or two entirely, making his abbreviated five to six-hoot call seem more like a casual "Hi." During these friendly greeting hoots he always averts his eyes. Finding out the small details of how an owl hoots may not be of great scientific import; but to me it is an exhilarating experience to discover even tiny details of nature.

Normally one hears great horned owls hooting in the wild only in winter and early spring, in the nesting season. Most of the other birds that sing also do so in the nesting season when they are sexually mature, and singing is a sign of sexual maturity. However, a caged male great horned owl at the Raptor Center who had been captive for seven

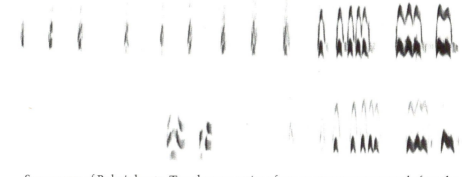

Sonograms of Bubo's hoots. Top shows a series of ten contentment grunts before the hoots. Bottom shows author's voice calling "Bubo, Bubo!" followed by Bubo's hoots preceded by only two low grunts. (Several notes that are distinguishable by the sonograms sound as one.)

years *still* had not hooted, though he was surely old enough. Bubo, on the other hand, hooted at least twice while he was still a nestling, and now after one year he gives hoots at any time on any day, provided I or anyone else comes near him.

Like the hermit thrush's melodious song, hooting in owls has at least a dual function. It tells bachelor males, "Keep out. This real estate is taken." To passing females (if the male is not mated) it means, "Come here. I've got some nice turf to raise a family on." Undoubtedly it is not quite so simple. Female hermit thrushes don't sing, but female great horned owls hoot. Do unmated females listen for the "marital duets" to determine whether or not the territory-holder is single, or is the hooting also an auditory signature that is necessary in these strong predators to prevent inadvertent injury by a mate in twilight, when visual cues are not reliable to differentiate friend from foe?

It is the many varied soft and hushed sounds that Bubo makes that I find most fascinating. I hear them only when I am next to him; they are his private sounds, reserved for intimacies. Whenever he comes close to me he gives one or a series of quick little nasal, reedy chuckles. His belly jiggles with each one, and they shake his whole body. In the context he makes them, they are like the purr of a cat or the babble of a baby. And like a babble, these little chuckles vary in tone and inflection. Sometimes they are slow and throaty, with an inflection at the end: "Hmm? Hmm?" When he is slightly irritated, the chuckles may grade into cackles of a higher pitch. It is these intimate details that bond friendship and promote empathy and understanding, and you learn such things from wild animals by living with them.

At 6:30 P.M. Bubo becomes active, having taken only a half interest in my bread making. To make bread I empty a package of dry yeast into some warm water, mix in a lot of whole-wheat flour and a little salt, let the dough rise, add some more flour, and then bake it. We don't have an oven, but a frypan on the stove covered with a dishpan that acts like a reflector oven is quite adequate. When the loaves are done they have real substance to them. They put mushy white bread to shame. This is my opinion, and Bubo's. He eats my bread with gusto, but totally refuses the commercial varieties. I reluctantly acknowledge, though, that Bubo's endorsement of my taste is probably a two-edged sword.

Having watched me mix bread dough, Bubo leaves the cabin to go back into his enclosure. Now, at 8:15 P.M., he is still exactly on the same perch as when he entered ninety minutes ago. He perches hunched down, his breast feathers slightly puffed out and drooping down to his toes. His ear tufts have a rakish tilt to the sides as he faces, as usual, in the direction of the hermit thrush's evening serenade. Occasionally his head bobs slightly up and down. A white-throated sparrow's flutelike notes drift up from the pines, and Bubo's head turns. One by one the other birds—ovenbirds, juncoes, and yellow-rumped and magnolia warblers—blend in their voices. Bubo watches (or listens) attentively. I too, strain my eyes and ears, but I do not see any of the birds. They are hidden in the now dark spruces, firs, and pines, among the stems of the white birches.

In the evening light the blue of the violets beneath the birches does not show. But the white four-petaled flowers of bunchberries shine brightly against the quickly darkening verdure of moss. Patches of fern, spirea, and goldenrod in the clearing are still clothed in luminescent green by the evening light, and over and through them flutter small white moths. The salmon-colored sky continues to darken over the ridge beyond the lake.

8:50 P.M. Almost dark. The hermit thrush still sings. A robin gives alarm calls. The creatures of the night are becoming active. Bubo stirs from his perch.

ALERTNESS in birds, as well as in humans, is correlated with daily cycles of body temperature. Our body temperature increases in the day and declines at night, as it does in small diurnal perching birds. One might expect the reverse in an owl, if it were truly nocturnal. Interestingly, however, the great horned owl also shows the lowest (about 1°C lower) body temperature at night. Body temperature drops about 1°C at sunset, and it has a sudden 1°C rise at sunrise (Siegfried, Abraham, and Kuechle, 1976). This cycle is roughly in accord with Bubo's behavior: he becomes active at sunrise, remains semiactive in the day, and sleeps in the middle of the night. Like his visual systems, these observations point to an evolutionary history of a bird who was originally day-active, and who is now adapted or adapting into a crepuscular and/or nocturnal creature.

A drop in body temperature of an inactive animal is often a physiological adjustment that functions to save energy. The less energy there is available, the more the body temperature should predictably drop. In a recent comparative study between red-tailed hawks and great horned owls, the fasted hawks lowered their core body temperature on the average 3.2°C at night. Great horned owls, on the other hand, showed relatively little amplitude in body-temperature fluctuations, and body temperature was not influenced by cold or fasting (Chaplin, Diesel, and Kasparie, 1984).

The reader may wonder how one can continuously monitor the internal temperature in a sample of red-tailed hawks and great horned owls. They are force-fed with 14-gram radiotransmitters whose beep frequency varies with temperature. The birds are kept close by (tethered by jesses on a 1.5-meter leash attached to a perch in a cage), and the beep frequency from the transmitter is monitored, preferably before the bird regurgitates it in a pellet. Also, it helps to keep your fingers tightly crossed in the hope that the birds' confinement does not affect their activity cycles and body temperature, and that stomach and crop temperatures are equivalent.

JUNE 7

The two crows, Thor and Theo, have meanwhile developed rapidly on a diet of cat food supplemented with scrambled eggs and home-made bread that I have chewed and moistenend for them. And, since Bubo has not yet been outside the cabin, they have been safe from him up in their nest in the ash tree. They have now long been out of their nest, however, and Bubo eyes them all too keenly from inside the cabin when they fly by. He butts against the windowpanes in his eagerness to get at them. But I will wait until they are still older and even better flyers before I let Bubo loose on them (or vice versa).

In the meantime he has other diversions. Today it is a bumblebee. The bee is rendered innocuous after I remove her stinger. But after I place this small fuzzy creature onto the floor, she promptly flips onto her back as though she still had her stinger, putting herself in a position to make good use of it. Bubo eyes the insect from 2 meters above. His pupils close and dilate in a flash as he gauges the distance. Will he pounce? The bee buzzes. Bubo leaps in terror from his perch without as much as a hiss, a clack, or a spread of feathers. In his great haste to escape he smashes against a window (which he has never done before) just above the sink, fluttering there like a frantic moth.

A blue-bottle fly who has been dining in the sink is disturbed and joins Bubo at the window. It hums, but Bubo pays no attention. The fly hums some more and, in its futile smashing against the invisible obstacle, falls down into a dirty bowl coated with remnants of corn oil. The fly, its wings now well lubricated, is undaunted. It flies back up to the window, but its flight tone is now considerably lower because its wing-beat frequency has decreased because of its heavier wings. It hums no longer; it buzzes instead. And Bubo now also flees from this "instant bumblebee," flying up to the rafters, from where he proceeds to the loft, from where he smashes against the window at the opposite end of the cabin.

Bubo eats insects avidly. Moths, beetles, grasshoppers—all are taken. Why does he flee from a bumblebee and a fly that sounds like a bumblebee? I crush the bee so that it no longer buzzes, and hand it to him. He grasps it now unhesitatingly in his bill, crunches it, and swallows it with gusto, signaling his approval with bill-smacking and soft cooing noices.

Perhaps Bubo has had a painful experience with a bumblebee in the past. But if he did, he identified the bee by what it sounded like, not by what it looked like. We can deduce that sound has also been important to innumerable other predators of flies in the evolutionary past, because flies that look like bees also sound like bees. By the same kind of logic it was correctly surmised that pollinating insects have color vision, long before this was proved by elaborate behavioral and physiological experiments.

I am surprised at Bubo's ability to distinguish the difference between the buzz of a bee and the hum of a fly, and to fear the bee more than the fly. Nocturnal owls should be little endangered by bees. However, there are many flies who parasitize owls, and these are indeed enemies of owls, and to fear *them* makes sense. (I would guess, however, that these flies are not noisy flyers.)

THE great horned owl, a large and powerful predator, is a safe haven for all sorts of parasitic vermin. No wonder many have chosen to live in and on it. The owl has more to fear from these small potential enemies than from large ones. Owls are the hosts of many parasitic insects, including bird lice (Mallophaga), fleas (Siphonaptera), and flies (Diptera) (Woods, 1971). The hipposcid louse-fly parasite of the

A louse fly taken from a barred owl in December.

great horned owl may in some areas not only drink its blood, but it may also transmit malaria (Herms and Kadner, 1937).

About the size of a housefly but greatly flattened dorso-ventrally as if it had been stepped on, a louse fly can crawl down through fur or feathers and hug its host's skin. Inhabiting a constantly warm environment, with all the food it wants directly at its feet, the fly must be about as near to paradise as it is possible for a fly to get. Except for a flightless fly (*Chionea* sp.) that walks on the snow, they are the only flies, as far as I know, who are active throughout the northern winter. A recently frozen barred owl, which had been killed along the highway in Burlington, Vermont, in mid-December, and which I dissected to sketch its leg musculature, had one of these flies drop out of its feathers onto the table. The fly was dead, not having survived its recent chilling.

Consulting an entomology textbook, I learned that it was most likely *Lynchia americana*, a species commonly found on owls but also reputedly occurring on "hawks and ruffed grouse." I wondered if the apparently wide range of hosts is due to its being more indiscriminate than most other ectoparasites, or because entomologists have not differentiated between species that may look alike to our eyes but are behaviorally and ecologically distinct. One species, the sheep "tick," is a wingless louse fly, and it only lives on sheep. Other species of hippoboscid ectoparasites are very specific to doves, deer, and other animals.

Near Christmas 1985 I thought I might get another chance to check for hippoboscid flies in an owl, but this time the story had a different ending. I was coming back from Maine on a busy highway at night when I noticed an injured or run-over bird near the center of the road. I halted quickly and snatched it up out of the way of oncoming traffic. To my surprise it was another barred owl, and more surprising, although there was blood oozing out of one eye, the bird was still alive. I put it on the floor by the back seat, and drove on, hoping for the slim possibility that it might recover. It did. And soon, too. After about fifteen minutes there was a commotion in the back, and then the owl appeared, for all intents and purposes, as much alive and active as one might expect any owl to be who is cooped up in a car and wants desperately to get out. The owl flew wildly back and forth against the side windows, the back window, and the front window. Then it nonchalantly perched in front of me on the steering wheel, just as I was entering the city limits of St. Johnsbury. I was not about to pluck it off with my bare hands, either, so there it would sit until it found a better perch.

This owl eventually settled on the back rest of the empty passenger seat beside me. I could there admire its beautiful big black-brown eyes and yellow beak tucked in between the two huge grey facial discs. As I kept glancing over and appreciating its seemingly friendly face, I talked to it, and the bird calmed down. I had been anxiously looking for the nearest patch of woods to release the owl, but now the bird seemed thoroughly at ease. It responded to my talk with little contented noises, just like Bubo's. It smacked its beak, looked at the passing traffic, and gave no indication of wanting to be elsewhere. This was turning into an interesting experience, so I let the owl keep its perch and drove on. Gradually its lids closed and it dozed off, only to become fully alert at each stop sign. The winding and bumpy country road through the woods to our house did not seem to faze it. Perhaps it felt it was in some particularly choice warm spot under some thick branches on a windy night. When we got home, the owl awoke, looked nonchalantly around, and flew swiftly out the open door into the forest. I have no ready explanation for the owl's bizarre tameness. I suspect that once it was inside the car, the bird lost touch with its world, and then perhaps lost all reference to the reality of its daily existence upon which its fear, and other essentials of survival, is based.

Some fly and beetle larvae live harmlessly in owl nests (Ryckman,

1953; Lee and Ryckman, 1954). But other fly larvae, maggots, burrow and feed directly in the owl's living flesh (Tirrell, 1978). Some of its protozoan parasites derive their names from the owl, bearing names such as *bubonis* (Cawthorn and Stockdale, 1981). Newly discovered mites (living in feather quills) have been named *"Bubophilus"* (Bubo-loving) (Phillips and Norton, 1978). Aside from these organisms, owls also have their share of worms; there are specialized tapeworms (Rausch, 1948) and over a dozen species of helminth worms in great horned owls alone (Ramalingam and Samuels, 1978). An owl can be a whole little ecosystem unto itself. I never did find out if the barred owl I picked up carried any louse flies.

One way many birds are thought to rid themselves of external parasites is by the use of ants in a special behavior called "anting," which I saw routinely in the two crows. Although the crows ignore individual ants, groups of ants attract them like a magnet. The more ants, the more irresistible they become—not to eat but to sit on! You merely need to scratch the top of an ant mound, and as soon as a half dozen or so ants are scurrying about, the crows drop whatever they have been doing and come. With their feathers ruffled out and their wings partially spread, they lower their rumps down into the ants. Undoubtedly this rude attention alerts the ants' nestmates, who come scurrying out of the mound to defend it. Soon there are squads of ants running up the feathers of the tail and wings. Ants clamp on with their pincers and squirt formic acid, their defensive spray. The fumes alert still more ants, and reinforcements come hurriedly. Meanwhile, the crows act as if they are intoxicated by the insects' undivided attention. They push themselves even more tightly onto the mound, drag their wings over it, and sometimes roll onto their sides and back, kicking their feet up in the air. They yank out some of the ants that are becoming embedded in their feathers, but more climb on all the time.

Although the crows act drunkenly, they do not appear to be having a lot of fun. They make angry cawing noises, and when both of them are on the same ant mound they peck angrily at each other, as if one is blaming the other for the misery it is suffering. Nevertheless, when I leave the area they are loath to follow. When they finally do come I feel a bit cruel showing them the ants at the next mound, to see if they will repeat the process. They usually do.

It has been suggested that anting behavior of birds is triggered by the smell of formic acid, an alarm pheromone of ants. Other stringent chemicals, such as alcohol, are thought to have the same effect. Do the birds simply get drunk from the smell? If so, then they can sober up more quickly than any human. The crows often converge on an ant heap simultaneously from two directions, so they could not be following an odor trail. They *look* at the scurrying ants, and the anting behavior follows instantly, so it is the sight of the ants that triggers the behavior. As to alcohol, I immersed cotton balls in both rubbing alcohol (isopropyl) and vodka (ethanol). The crows eagerly take the cotton, as they take any strange or conspicuous object. But then they quickly drop it, sneezing in irritation to both alcohols. They do not squat down for some anticipated other sensations.

LATER on in the summer I had the opportunity to test Bubo on ants, also. He watched me scratch the ant mound open. Plenty of ants were scurrying about. However, he paid no attention to them, and if he did see them, he had no compulsion to lower his rump into their midst. I dropped a piece of meat onto the mound to lure him in. As I expected, it worked like a charm, but Bubo was not enamored by the ants. He walked tall and stiff-legged as ants quickly ascended his legs while he tried to pick them off with his bill. Moisture started to run from his nares, and he shook his head and flew off. Anting may be fine for crows, but Bubo would have none of it. Ants make him sneeze.

JUNE 8

It is over 32°C (90°F) today. Bubo is perching up in the loft, standing tall and erect, and exposing his surprisingly long legs to cool himself down. (Normally his legs are totally invisible when he perches.) His wings droop to his sides until they touch the perch so that his thinly feathered sides and underwings are also open to the air. In addition, he sleeks down his feathers so that he looks skinny.

He looks as though he has a cold: a thin stream of fluid runs from his nares and drips off the end of his bill; occasionally he sneezes. His nasal drip, however, is related to the heat and his weak kidneys. When he evaporates water from the mouth and throat to cool himself off, the salt concentration of the blood increases, and the salt has to be voided. Birds cannot concentrate salt in their urine, and many species (including sea gulls and raptors) have salt-secreting glands that dump the excess salt in a thin solution from the nose (Cade and Greenwald, 1966), much like turtles' or crocodiles' excess salt is excreted in "tears" in secretions voided into the orbits of the eye.

Bubo on the rafters inside Kaflunk. His posture shows that he is hot and is trying to cool himself.

June 9

This morning, after coming to my call and being fed their cat food, both crows fly up to the ridge of the roof. They bask there in the morning sunshine, until the peacefulness is shattered by their raucous screaming. Out of the corner of my eye I see them fluttering off, and I also glimpse the large black shape of a raven. The raven makes a hasty retreat as I run out. One of the crows is gone, but it later emerges out of the underbrush. The raven comes back later, circling low over Kaflunk. The crows, perched silently in the branches of the birch, observe him closely. Crows vehemently mob ravens in the spring during nesting season, just as many other smaller birds vehemently mob crows. But in my fairly extensive observations of ravens and crows over two winters, I have never seen crows mob ravens even near animal carcasses where both were feeding, and vice versa. In the winter, however, crows did not seem to diminish their mobbing of owls and red-tailed hawks. Both corvids could steal each other's eggs. But the hawks could be a danger at any time to the adults as well as the young. Does mobbing serve very specific functions that vary seasonally?

June 10

I'm ready for bed early because I get up before 5 A.M. every morning, adhering to Bubo's strict schedule. But Bubo's schedule sometimes demands even greater sacrifices. Not long after dark, as I am trying to sleep, Bubo taps on the window. I try to ignore him, of course, but the tapping continues irritatingly at intervals of two or three minutes. The tapping becomes louder, but I think I can sleep through it, until a concern keeps me awake: the ancient windowpanes are thin. Should I let him stay outside and risk his cracking the window, or let him in and risk my sleep? I opt for the latter choice, rationalizing that he may settle on a rafter and sleep, since he usually does sleep most of the night in his cage. I am not getting much sleep anyway.

Bubo, who has been flying about in the cage for the last half hour, now settles peacefully on the back of the chair next to my bed. Relief! Except that it is another hot night—nearly 32°C. I lie bare on top of the covers, listening to the evening chorus of the mosquitoes. They can be heard from afar as they search around methodically in the dark. When an individual begins to home in on its target (me) and starts to pay particularly close tribute, I wait patiently for her (male mosquitoes don't bite) to land to terminate her mission. Luckily these bites are not dangerous. The mosquitoes here are quite civilized and do not transmit malaria, yellow fever, encephalitis, or any of dozens of other diseases.

Bubo, meanwhile, is also hot. Through the drone of the mosquitoes, his gular flutter sounds like a butterfly beating papery wings on the windowpane.

He is probably less bothered by the mosquitoes than I am. A thick coat of feathers functions better than any screen. Blackflies, when they hover about him during the day, occasionally bite him on the eyelids. But his body is free of them, as well as of the mosquitoes because of his protective shield of feathers. Owls and hawks may need this protection because they sit perched still for long periods of time, enabling flies to gather. However, Bubo now has Achilles' heels—the backs of his legs where the jesses have worn off the feathers. Mosquitoes and blackflies gorge themselves there into rotund purple blobs, and fly off replete and unscathed.

BITING flies are most prevalent in the north (although I once encountered my fair share in a Florida salt marsh), and some northern owls can become quite anemic from fly bites, according to Katherine McKeever at the Owl Rehabilitation Research Foundation in Ontario, Canada (personal communcation). Could the furry legs and feet of northern owls have a function besides insulation? It is hard to know how these northern owls would fare if they did *not* have feathered legs, but the small bare patches on Bubo's legs give a good hint. Feathers are a remarkable evolutionary invention that has been put to many uses. What other adaptation could simultaneously serve as a sunshield, insulation, mechanical armor, a raincoat, a device to make flight possible, decoration for sexual signaling, camouflage, and possibly a guard against bloodsucking insects and the deadly parasites they sometimes transmit? Sand grouse in Africa even use their belly feathers as a sponge to carry water to their young.

Biology teaches one broadmindedness, because often a number of entirely different hypotheses are simultaneously true, to varying degrees, in different species. Each species is different, because each is better at occupying a specific niche than another one. And it is precisely because each is a unique adaptation that makes it possible for us to see patterns and to understand in what ways and why we are all alike.

MY REVERIE is terminated with a "splat," and I visualize other things. There is the flutter of wingbeats. A mysterious clank. A little swish. Rattling silverware. Nothing serious. I try to block out the sounds, but I seem to be listening, nevertheless. A long silence. "Click, clack, click . . ."—he is walking across the floor. Splat. Another flutter. "Rip." Silence. "Rip." Silence. "Rip." Now I sit up. Bubo is on my table and the beam of my flashlight catches him holding my favorite shirt—what is left of it. I jump up and yank it out from under him. Bubo chitters indignantly.

All is quiet for a few minutes after I crash back onto the bed. I am getting drowsy again, but the tap of his toes marching on the floor does not sound reassuring. A flutter. He makes a soft landing at the end of the bed, close to my toes—too close. Visions of his massaging my naked thighs keep me awake. Enough! I jerk out of bed. End of experiment. I won't spend the night in the same room with a great horned owl. Period.

I open the window and reach for "the glove." I can't waste time. The mosquitoes, as always, do not. I hold the glove near Bubo's toes so he can hop on and I can maneuver him out the window, but I get no cooperation at all. He chitters, and the chittering picks up in volume like an engine revving up. He is angry and snaps at the glove; he remembered that I had once used it to hold him forcibly on my fist. Like the fox who condemns the trap, not the trapper, Bubo blames the glove, not me. As I shove the glove under him he applies his death grip to it, bites down hard, screams, and then hops off. This won't do. I grab a chair and hold him at bay, maneuvering him toward the window. He first holds his ground and attacks the chair, then flies over it to the opposite end of the room. It is a contest now, one I am determined to win. I am well motivated to get him out, knowing quite well that he will wake me up at 4:30 A.M., not a very long time from now. I eventually succeed in slamming the window down behind a hissing, bill-snapping, biting, clawing fiend in the dark.

June 11

Like an alarm clock, Bubo wakes me at 4:34 A.M. by drumming on the window beside my ear. He joins me for breakfast, sharing some of my pancake; I make it by mixing lots of eggs, some condensed milk, and a little water and salt with whole-wheat flour. He likes my pancakes either with or without Maine maple syrup. He swallows a mouse for a treat, but I opt for toast. He hops onto the back of my chair, making his friendly grunts while I caress his head, and he nibbles on my fingers endlessly. After I have had enough of the morning session of touch-and-feel, I try to write, but Bubo keeps inserting himself between me and my pencil. He wants fingers to nibble on, and fingers he gets. I do not disagree with a great horned owl unless it is absolutely necessary.

I can understand and identify with Bubo so thoroughly that I know why he is the way he is, and I expect him to be the way he is, even though I sometimes do not like his ways and will defend myself against him if I have to. My understanding transcends any feeling of blame or forgiveness. I can only accept facts about him and work within them to achieve our mutual well-being. Right now it seems my well-being is a little less than optimal.

After breakfast I check, as usual, under the bed, and retrieve two dead woodland jumping mice that had been deposited by Bunny. As I drag one across the floor on a string, Bubo dives from the rafters and nails it in an instant. It is only a short diversion. A roll of toilet paper—always a personal favorite of his—occupies him longer, although he can reduce it to tatters in a remarkably short time. A small carabid ground beetle scurries across the floor. Bubo stares at it for about a minute, then runs to it and eats it. I offer him a raw clam. Like last year, he still refuses raw clams, and I am not ready to steam or fry them to his order. Instead, I peel a cucumber. He watches, and when I am done he hops onto the table and grabs the long peelings. He methodically reduces each peeling to little bits that are flicked sideways with a quick shake of the head. He swallows about every tenth piece he bites off. Cucumbers are not his thing.

June 12

Bubo looks out the window at the crows who now fly around the outside of the cabin every day. After watching their now sterling flight performance, he runs across the floor and stabs violently at a piece of cardboard. Frustration? He looks around some more, and in one or two violent leaps strikes feet first on the leather glove, holding it with his left foot and repeatedly stabbing it with his right—pounding down hard with extended talons and locking them in a shearing snap. Then he lands on the back of my chair and softly nibbles on my ear.

Bubo is in a frisky mood, and so am I. For fun I will see if he still remembers toads from last year. I release a toad from a coffee can under my desk. It takes one hop onto the floor, and Bubo hesitates only briefly before he pounces on it. He lowers his bill and starts testing his prey with eyes closed. The reaction sets in: his grip relaxes, he steps back, his eyes open wide, and he studies it suspiciously. Toad hops away, leaving the owl violently shaking his head. The head shaking gets more vigorous, and an indelicate spray starts flying out of his bill. He retches, and then regurgitates one of his own feathers, which he had pulled out while preening and swallowed only minutes ago. It is now a soggy spit ball. He shakes his head some more, and his nares continue to drip.

Pulling himself together, Bubo decides the toad deserves a closer look. He walks up to it again, stares at it some more, and cautiously circles it, all the while staying at least 10 centimeters away. Then he walks to the cat's water dish and drinks until it is drained. Bubo has just mimicked my own reaction to my least favorite of condiments, hot peppers. I conclude he has forgotten about last year's toad.

The ability to remember that toads are noxious critters would seem to be a highly adaptive trait. But might too much remembering also be harmful? *We* know that all toads are noxious, but it is probably not adaptive that one trial would turn the owl off toads forever. Not all toads look exactly alike, and Bubo has no way of knowing beforehand whether *everything* that looks similar to a toad is noxious, or whether for every noxious toad there are one hundred toads (or close mimics of them) who are good to eat. If his first experience happened to be with a noxious toad, and his memory were perfect, then he could be shutting himself out of a good food supply. It is better not to remember some things for too long. Forgetting, too, is adaptive. It keeps one's options open, which is especially valuable in a changing environment.

Bubo will not have enough time in his life to test every single object for palatability. He will be forced to take shortcuts, to generalize, to extract relevant stimuli that differentiate potential prey from non-prey. As Samuel Butler said, "Life is the art of drawing sufficient conclusions from insufficient evidence." Forgetting, and being willing to make a few mistakes, is part of the game.

JUNE 13

The crows are walking on the granite ledges just outside the door. Suddenly I hear their excited alarm, "Caw—caw—caw—," and they fly up, zoom around the corner, and land on the roof. Then Bunny nonchalantly makes his appearance. The crows have never had an encounter with the cat before. But they make alarm calls from the start. Good. They will likely be watchful of Bubo also. I can now release the owl, because the crows are not only acrobatic flyers, but they have also demonstrated perhaps the most useful basic and universal law of survival—fear.

Owl vs. Crows and
Jays vs. Owl

WILL the crows mob the owl? Or will the owl try to hunt them, and if so, what techniques will he use? My adrenaline is up. I am near the critical point of a project I have been setting up for over a year. And it is a rather unusual project at that. Naive owl will now meet naive crows, each for the first time. Anything can happen.

Tomorrow morning, when I let Bubo outside for the first time, he will not have been fed for two days. I hope he will be hungry enough to stay close to me and the crows, rather than wander into the forest immediately.

JUNE 14

As always, Bubo knocks on the window at 4:30 A.M. After inviting him into the cabin, I open the door to the outside, luring him out with a dead flicker saved from my jog yesterday afternoon. The lure, how-ever, is not necessary. Bubo waddles straight out the door and turns around to examine the cabin's exterior and the surrounding trees in some excitement, totally uninterested in my dead bird. He wants a crow.

The crows, who have not yet been fed this morning, come to me im-mediately. Strangely, they pay no attention to Bubo. Do they have to learn to avoid owls, while innately knowing the danger of house cats? Their lesson begins.

Bubo immediately takes off, chasing Theo (the smaller, presumably female bird) across the clearing and into the woods. Then he returns and lands on top of Kaflunk. Maybe it won't be as easy as he thought. No matter, there are other things to do. Now he sees the rock pool in the ledges where he bathed last year. The crows can wait. A bath, ap-parently, cannot. He takes a long leisurely dousing. The crows glance

Bubo before, during, and after his bath.

over occasionally to watch him splash, but they are only mildly interested, if not downright nonchalant.

Thoroughly soaked and waterlogged, Bubo flies onto his favorite old perch of last year, in the white birch in front of the cabin, and from there he now begins to take an earnest interest in the crows.

He has always preened after taking a bath. Not today. Still dripping wet he stands stock still and glares at the two crows that have just been fed. I doubt if he realizes it, but he has already made a great tactical error. A waterlogged owl will not be a graceful flyer, nor a noiseless one.

Theo watches Bubo from the roof. Thor, beside her, is less alert. He preens himself. Bubo ignores Theo after giving her only a glance and fixes his eyes on Thor instead. Thor continues to preen, and Bubo launches himself off the birch directly at him. But he takes off in the nick of time. Both crows fly off, circle around once, and return to the roof. When Thor resumes preening, Bubo attacks him again.

Bubo remains so absorbed in the crows that he has forgotten all about his own preening. Instead, he makes his way to a higher vantage point in the birch. His head bobs up and down in excitement as he again gets his bearings on the crows who are nonchalantly perched on the ridge of the roof. Eventually Thor turns his back to him, and within seconds, gravity-assisted by his wet, weighted feathers, Bubo plummets down after him. He step daintily aside. Missed again. Bubo, it seems, cannot or will not alter his flight trajectory after it has been set.

Bubo is not having any luck, but it is evident he is still excited because of the tiny droppings he excretes at frequent intervals. Perhaps he'll eventually eat crow. But just in case he won't, he eats his flicker instead. This may be his second mistake.

After his meal Bubo reassumes his perch on the birch while the crows cavort on the roof. Theo watches owl. Owl ignores her but again watches Thor, who seems as unconcerned as before. The second he turns his back to Bubo, he plummets toward him. The result is a complete repeat. Still no crow in the talons. The crows now land on the ground near me, waddling along as unconcerned and self-assured in their unassailability as the devout on their way to church. Theo stops to pick at some pine needles. Bubo, who has not yet taken his eyes of the crows, drops toward her. But as he gets close, she looks up, and Bubo veers off, aborting his swoop. Theo does not even have to hop

Bubo hiding in thick spruce branches.

aside: she barely breaks her stride. Doesn't she know she is in danger? Maybe she *is* blessed.

Bubo now tries a new tack. He ascends to a higher perch—at least 25 meters up into the large red spruce in our clearing. Up there he is hidden from sight in the topmost thick branches, and only his yellow eyes can be seen peering from underneath some boughs. Although he is well hidden there from the crows, he has the disadvantage of having a very long attack flight, making surprise more difficult because the crows constantly move their heads in different directions. They are bound to see him sometime during his descent. And if they do, they can easily avert his charge. For the time being, though, Bubo just watches, and the crows are at ease again.

Theo later returns to the top of the cabin and walks seemingly unconcerned toward Thor. Bubo uses the opportunity to pop out from behind the twigs in the spruce, and he descends like a bolt. But again Theo sees him coming, looks up, and crouches, ready to spring out of the way. One look that shows him he has been seen is enough. Bubo

tries no further. He veers off in midflight, and she seems hardly ruffled; she hops about on the roof a bit more, and then descends to the
woodpile. Bubo, meanwhile, flies onto the roof and looks directly
down on her. She waddles up to me and begs for food. Bubo again
seizes her unguarded moment and pounces down. Again he does not
follow through when he is aware that Theo sees him coming, and he
flies up into the spruce again.

Theo goes to the mountain ash, and when her back is turned, the
persistent owl attacks from the spruce. This time she is finally surprised, but she is saved by a few twigs in Bubo's way. Theo leaves, silently, and Bubo flies back to the spruce. Meanwhile, Thor has returned to the roof, and Bubo swoops down at him. Thor caws loudly
at his approach, and noting that his surprise is destroyed, Bubo just
swoops on by. Thor does not even leave his perch.

Both crows are playing on the roof, picking at the old cedar singles.
Bubo now tries a perch almost directly above the cabin roof, on a high,
exposed limb of the birch. They merely hop aside as Bubo dives at
them, but then he chases after Thor on foot, who flies away, finally
cawing. Theo hops down to the woodpile, and starts to preen. The
crows do not seem to be getting any wiser to the owl. Or are they?

Bubo is now starting to dry off, and he does not sound quite so
heavy-winged any more. He makes another pass at Thor on the
ground below him. This time Thor seems completely taken by surprise, and sees the owl only at the last instant. He takes off cawing.
Well, the crows are beginning to show at least *some* alarm.

The blue jays, who have a nest with young nearby, have not sat idly
throughout all this activity around Kaflunk. Indeed, their activity has
been quite frenetic the whole time. One or another of the pair has continually been maneuvering to situate itself behind and directly above
Bubo, much as Bubo jockeys for position relative to the crows. Then,
while Bubo watches a crow, the jay plummets down from behind,
grazing the back of his head and back. The jays also scream at the owl
constantly, and they hop directly in front of him, but no matter how
close or noisy they are Bubo never shows the slightest indication of
being tempted to chase one. Since the jays swoop at him from behind,
from a range of 3 meters or less, they usually get to his head before he
has a chance to turn it. And he *always* turns it after they swoop at
him. Do the jays keep him from looking for the nest? I counted forty-
seven swoops, and every one of them was initiated when the owl was

Bubo drenched by rain. The lower photograph shows him blinking by drawing his nictitating membranes over his eyes.

not looking at the attacker. It was as if the jays were insisting that he keep his eyes on *them*. In twenty-two of the swoops, however, Bubo turned toward the attacker before it was within 1.5 meters, and in each of these instances the jay invariably did not follow through but broke off the attack. I was amazed at how he ignored the frenetic adult jays prancing directly in front of him, while concentrating his considerable attention on the more inexperienced crows a long distance away. Both attackers—the jays and the owl—paid meticulous attention to the movements of the head and eyes of the object of their attacks.

6:45 A.M. It has started to rain, and after two hours of nearly constant activity all the birds are taking a break. Bubo goes to rest under the thick branches of the spruce. The jays have stopped scolding, and the crows are asleep in the birch tree.

8:45 A.M. It is still raining, and Bubo has taken a second bath in the pool. Apparently one year is a long time without a bath, so now two in one morning, plus frequent drenching by the rain, feels about right. After the bath he perches in the pouring rain on top of the roof. The crows are still less than 12 meters away in the birch, and they are now singing. First one, then the other, and sometimes both of them together warble like hoarse mockingbirds for ten to fifteen minutes at a stretch. They preen, stretch, and shake off the rain that soon soaks them again. Their serenade sounds pleasant in the light rain, but when the rain becomes heavy they stop singing and pull their heads down onto their shoulders.

Bubo does not take his eyes off the crows. He shakes his heavy wings weakly, as if hefting them. But he is no longer as ready to attack as before. His confidence is diminished, perhaps because of his wet wings or because of the past failures.

Theo stops singing and hops to the ground directly below Bubo, who continues to move his wings weakly at the shoulders, staring at her, perhaps weighing the chances of a successful pounce. He decides to give it another try. Down he drops, just as Theo looks up. Bad timing. Bubo again flaps off in defeat like an ungainly vulture. But if success is going from failure to failure without loss of enthusiasm, as Winston Churchill claimed, then Bubo is still successful.

Meanwhile, the crows have returned to one of Bubo's favorite perches, the roof. He now has decided to come back, too, and lands 2 to 3 meters from the crows, but he chooses to ignore them. They, too, are now either unconcerned or else are putting on a good act of being composed. Chalk one up for the crows—they seem to have won the battle so far. But with two baths and a heavy breakfast, Bubo had been at a disadvantage. I hope the crows are not now lulled into a false sense of security.

12:47 P.M. Both of the crows and Bubo are perched in the birch. The crows are 3 meters across from him, quite at ease. He flies over, landing about half a meter in front of Thor. I expect him to take off, but he does not. Instead, he opens his bill, caws loudly, puffs out his feathers, lowers his wings and head, and gesticulates with his head as if about to deliver a telling jab. Bubo acts surprised by this energetic display— and he just stares. But then as Thor's act gradually subsides in its frantic vigor, Bubo shifts his weight from one foot to the other and advances one step closer. Thor resumes his act. Every time Bubo moves, Thor responds with a new bout of cawing and posturing, which keeps Bubo from proceeding further with his plans. It is an impasse. After a decent interval (about two minutes) Thor hops to another branch and loiters there, idly picking at leaves as if he were totally unconcerned about a mere great horned owl, and then he flies into the woods.

1:30 P.M. The sun is out and the birds are finally completely dry. Bubo is on the birch, and the crows are on the ground near the rock pool. Bubo is taking a renewed interest in them. He flies down onto the woodpile for a better vantage point, and the crows accommodate him, wandering back up toward him, cocking their heads. They see him. Still, Thor is edging up closer, stopping every once in a while to peck the ground, pick on twigs, pull on grass, and to look at him. As he gets nearer he hesitates more, extending his neck, a foot, then taking another hesitant step forward. Several steps. He stops, looks up again, picks up a stick, and twirls it idly in his bill. Drops it. Picks at a rock and advances a few more steps toward Bubo. Now he has worked his way almost directly under him. Bubo is not fooled. He seems to know he is being toyed with, and he soon ignores him and fixes his eyes at Theo, who is almost out of sight way down by the pool, basking in the sunshine with spread wings. Thor, still not getting Bubo's attention, now becomes even more brazen and flies up onto the woodpile, edging ever closer to Bubo but acting unconcerned, as though he no longer sees him. What is the basis of the crow's strange behavior? I doubt that he consciously knows, but it seems to me that if he can invite attack by making himself appear *as if* he is not aware of the owl, and still get away when attacked because he actually *is* aware, then perhaps Bubo might, after many more unfruitful attempts, not bother to attack Thor (or another crow) even when he is indeed off guard, which is eventually probably inevitable.

One crow is very close, but too alert to be caught, while the other one is probably not alert but far away. Bubo remains immobilized by indecision. With experience he will probably learn to evaluate his options, to distinguish the fake from the genuine inattention, and to act on the best choice.

JUNE 15

The crows already seem more skittish than yesterday, especially Theo. Nevertheless, they display some daring. Thor slowly walks along the ridge of the roof, directly in front of Bubo, who is now on the birch. He picks at shingles, snaps at mosquitoes, and gives Bubo occasional glances as he dawdles along, as if again daring him to attack. Bubo tries. Both crows then make a spectacular, fast, graceful flight, chasing each other around the clearing, while Bubo watches with a bored expression.

Bubo makes several more half-hearted chases after the crows and then leaves them alone. He soon finds other things to do. He now attacks the moss-covered hummock, pouncing on it with both feet, ripping out tufts of grass and bits of moss and tossing them right and left. Success, at last!

Later, as I watch the sun come up while sitting on the ledges, Bubo comes to me and hops onto my leg. For a half hour we nuzzle, tickle, and caress. He even walks up my arm, and his tread is soft. There is no pressure in his talons. I do not need a glove, and I decide not to use it any more. He is friendly now and he is getting friendlier every day. While I'm at it, I decide to get rid of the jesses he got at the Raptor Center. I hate the sight of them, dangling from him like prisoner's shackles. Snip, snip, off they go, forever.

Bubo joins me for a walk in the woods after breakfast. We have not gone more than 60 meters when I see a nest about 4 to 5 meters up in a small pine. I tap the tree, and to my surprise several young blue jays fly off clumsily with rapidly beating wings. Bubo, of course, sees them, too. In an instant he chases after the nearest one, and as I go after the hopping, fluttering jay, trying to save it, Bubo charges past me and catches it like a veteran. With the bird in his clutches he wheels about, facing me with outstretched wings and a loudly clacking bill. "Leave," he seems to be saying, "It is mine."

His instant resolve, energy, and expertise surprise me. He has never shown the slightest intention of chasing an adult jay, and this young one is already feathered out, and it can fly. Bubo had been triggered to attack by its *behavior*, as he had with the wounded squirrels last summer. Perhaps he only has a chance against prey who are disadvantaged because of age, weakness, inattention, or inexperience, and he knows it. He already seems to have an impressive and perhaps inherited grasp of his victims' details of behavior that might betray their vulnerability. However, the evolutionary game always works both ways, and I am willing to bet that potential victims have evolved countermeasures that sometimes help to save their hides, such as acting alert and vigorous, as best they can.

After he captures the young jay, the adults divebomb him and scream. Every time he turns his head or looks down, he is with virtual certainty divebombed from above. Within the hour I tally eighty-seven more swoops at the owl. Bubo tries to follow every swoop of the attacking jays with his eyes. When his attention wanders from those scolding him, they come closer, yell louder, and attack again. As the attacks continue, Bubo shows annoyance—he begins to hiss at his attackers. Eventually he departs to a more peaceful corner of the woods, perching close to the trunk in a dense red spruce where they can't get at him. I suppose he is exactly where *they* want him, because he is no longer in a position to find the other young.

Bubo did not get more youngsters of that nest. The jays' antics had one major effect—diversion. The jays' calling probably also served the twofold purpose of warning the young to be quiet and educating them. However, a few of those yells should have been sufficient to keep them quiet. But very vigorous activity was required to divert Bubo and get him away.

Having watched Thor and Theo and the mobbing jays interact with Bubo, I doubt that a great horned owl could catch an adult bird that is on its guard, and the owl would therefore not waste much energy in attempting to do so. Mobbing this owl, and perhaps others, is generally not dangerous for adult birds. Most likely those birds who become meals for owls are those who have not seen it. In contrast, I doubt that the same pair of jays (if by themselves) would risk such vigorous mobbing if their enemy were a Cooper's hawk, which would be at least if not even more likely to eat young jays. I did not have a tame accipitrine hawk to test this, but they are common near Kaflunk and I have not seen birds mob them. Nevertheless, many birds do routinely mob the slower Buteo hawks who mainly feed on rodents, reptiles, amphibians, and nestlings.

Bubo, safely backed into a dense tangle of twigs, has escaped the direct divebombing of the jays. Now he starts to nuzzle his prey, the young jay, and to take a few bites. However, he is not hungry, since he had had his fill of woodchuck meat before we went into the woods. A full stomach, however, had not eliminated his enthusiasm for the hunt. As with other predators, the hunt itself has its own rewards.

Defending the Meat
and the Manor

BUBO flies out of the woods, trailing his jay in his talons. He continues on across the moss-covered ledges with their scattered trees, and onto a patch of reindeer lichen to a low, stunted spruce. Here he drops his prize, examines it briefly, then picks it up in his bill and waddles full speed ahead. I watch from 30 meters away, and he stops, turns, and curiously looks at me a long time, then shoves his prize under some ground-hugging evergreen branches. Later he returns to Kaflunk.

Not wanting to waste good meat, I intend to retrieve the jay. But as I get near it Bubo swooshes over my head from behind, coming close enough to ruffle my hair. Strange. He has never flown so close to my head before. He lands in front of me, stares at me, and sleeks his feathers back. He is giving me a message, but I do not read it and continue walking around a curve and out of his sight. I am not, however, out of his mind. Apparently he has not forgotten the jay, and he knows very well where I am headed. I reach the spruce tree and am just ready to pick up the bird, when he comes swooping toward me, flying close to the ground. He lands and starts running at me, loudly clacking his bill, and then hooting. In an instant he has reached my legs and starts tearing, biting, and scratching at my loose pant legs. My heart is pounding at this totally unexpected maniacal fury, and I make a hasty retreat into the brush, with him in close pursuit. But he follows me only a short way and then runs back to retrieve the jay and hide it elsewhere. I have always been puzzled by reports of great horned owls attacking people when there was no known nest to defend. They are no longer so puzzling to me.

JUNE 16

Bubo is not a quitter when there is any chance of success. So far I have counted thirty-six unsuccessful crownapping attempts. But 95 percent of the winning is in the showing up, and today he was lucky. He swooped off the birch and neatly picked Theo off the roof. She exploded in a violent, vociferous uproar of thrashing wings, and was airborne again on her own power within 30 meters. Bubo pursued her into the woods but then landed on a branch as she dove into the underbrush. Thor followed them both and landed next to Bubo, possibly to distract him. Indeed, Bubo returned to the birch by Kaflunk, and Thor then flew onto the roof directly in front of him. Bubo then gave chase, and Thor led him once about the cabin, returning to the roof. Meanwhile, Theo remained silent and out of sight.

Two hours later, at 8:50 A.M., Theo is still nowhere around. Thor makes a flight high above the clearing. He has never flown high before. Is he looking for Theo? Later he, too, is missing, but Bubo is preening himself on the birch. I call for the crows. Thor comes from the forest and I feed him. Then I hear a faint fluttering from the thicket where he had just been, and that is where I find Theo perched close to the ground, dried blood on one leg and wing. She begs in a feeble, plaintive voice, and Thor comes back down into the thicket and perches directly beside her again.

IT TOOK Theo two days to recover from her wounds before she again showed herself in the clearing. Theo's suffering is probably mild in comparison to what often goes on routinely in nature. One can look away, but that does not change the reality. Bubo has to eat, and he will have a severe problem if he does not have the skill to catch live prey. I am tempted to provide live rabbits for him so that he can sharpen his predatory skills; but I admit to being squeamish, especially since tame rabbits will not have a chance to escape. Over two summers I presented Bubo with 73 birds (31 species) and 119 mammals (16 species). Almost all of the birds were road kills, and about 90 percent of the mammals were killed by the cat. If cars and the cat had not killed these animals, then Bubo (and I) would have had to do it.

Taken over an owl's lifetime, the numbers of animals it eats to keep alive appear to be huge. Does an owl have a large ecological impact? Suppose, for example, that an owl eats one squirrel per week (along with rabbits, songbirds, mice, and more). Squirrels are a favorite prey of great horned owls, so this is not an unreasonable assumption. In a year the owl could eat fifty-two squirrels, a potentially large impact on the squirrel population in a local area. But squirrels, in turn, raid bird nests, and if each of the fifty-two squirrels had destroyed one nest with four young per week for two months every summer, then the impact of the owls would theoretically be 1,664 *fewer* songbirds killed for that year. Nevertheless, areas containing an owl are not teeming with songbirds. The balance is hardly ever so simple or direct.

JUNE 17

Late in the morning I hear hooting, and then human voices. My nephew Charlie and his girlfriend Jody are coming up the path.

Bubo wastes no time making them feel unwelcome. By defending his territory he is defending not only food that he has cached, but also that which has not yet been captured. He attacks Charlie's shoes, and I cannot pry him loose even with a stick, but a bucketful of water does the trick nicely. Bubo, now soggy, resumes his perch on the birch, peers into the cabin, and hoots continuously at the company. He is

Bubo attacking visitors.

getting more angry, rather than less. Two hours later, every time they open the door he swoops down, and they duck back inside. How will they be able to leave? I find a solution: they need a hostage—Bunny. They must carry Bunny with them, because Bubo will not get near the cat. And so it happens: they run out the door, with Charlie clasping a meowing, protesting Bunny and Jody clutching Charlie, and Bubo in hot but not so close pursuit.

Hours later. It is high noon under a hot clear sky on a windless day in mid-June, and Bubo looks down the trail by which the company has departed, and he hoots for hours on end, one series of hoots after another, with no interruption. Until . . .

I hear a loud "swoosh—," and glancing out the window I see a hawk already receding in the distance. A few leaves within centimeters of Bubo's head are moving from a draft, evidence of the hawk's passage. The large bird had threaded its way through that tree at incredible speed, a feat that would be impossible for Bubo.

ONE HAWK, the red-tailed hawk, is the great horned owl's closest competitor (Hagar, 1957). Both hunt almost the same kinds of prey, and the two species do not get along well. Great horned owls take over nests of the red-tailed hawk, and sometimes they even capture and eat the adult hawks themselves. The hawks and owls segregate some-what because they breed at different times of the year and because they have different daily activity times (Springer and Kirkley, 1978). I doubt that Bubo felt chastised for being a competitor, but he stopped his hooting.

The hawk, however, might have known its enemy. Arthur C. Bent (1961) reports:

> In the middle of a bright day in April, while we were hunting for nests of the red-tailed hawk in the woods of Plymouth County, Mass., we saw a pair of these hawks sailing about over a large tract of pitch pine timber, half a mile or so distant. Half an hour or more elapsed before we began a systematic search for their nest, when only one of the hawks was seen, circling back and forth over the woods and evidently looking for something. We had not gone far into the pines before we saw a great horned owl fly from a small pitch pine; on closer inspection, we saw a great mass of feathers on a flat branch near the top of the tree; it was apparently the owl's feeding roost, as there were feathers and droppings on the ground

beneath. I climbed up to investigate it and was surprised to find the wing of an adult red-tailed hawk which had recently been torn from the body of the victim; the flesh was still fresh and warm. I had no doubt that the owl had just killed one of the hawks that we had seen sailing over the woods less than an hour before.

June 18

Theo has returned to the clearing. But she still acts lame and her voice is feeble. Bubo could probably catch her now, but Thor again acts as if he is deliberately trying to divert Bubo. Thor draws five attacks and after each one he turns his back on him and picks at the ground, nonchalantly, as if to invite still another attack. I find it miraculous that he always gets away when Bubo is almost on top of him. I did not suppose the crow's actions were deliberate, but I now think that this is a good possibility.

Eventually the crows retire into the dense woods, and Bubo remains half asleep on the birch in front of my window as I work at my desk. Then he studies the ground, and pounces. I rush out just in time to see him swallow a large, smooth caterpillar. I have never fed him caterpillars before, but he seems to have developed an eye for good things to eat—maybe anything that moves.

June 19

It poured most of the night. The crows are drenched this morning, but Bubo is quite dry. There is only one place where he could have kept dry all night—in our log cabin (with still open gable ends) down in the clearing. Apparently he remembered the place from last year, including how to get there. As before, he probably perched on one of the collar ties under the roof.

He swallows two red-backed voles, a shrew, and a Nashville warbler, and then he eyes Thor, who has come and perched directly in front of us to rearrange his wet black habiliments. He pushes one wing out to the side, picks at the shoulder, reaches under the wing, and behind it (Bubo ducks as if ready to launch himself). Thor's head is forward again (Bubo relaxes) and he draws breast feathers through his thick black bill. He stands on his right leg, pushes the left one acrobatically up between the left wing and body, and scratches the back of his head with one of his toes (Bubo tenses up). Thor reaches back to preen his tail (Bubo launches himself). He is almost upon Thor, when just in the nick of time he hops a few centimeters aside, as if unconcerned. Bubo, propelled by sheer weight and momentum, does not change course, and misses his mark, as usual.

If victory is mostly a matter of just trying, then it had to happen. Bubo has tried again, and again, and . . . eventually he got lucky and managed to snag Thor in his talons in broad daylight. But Thor put up a good fight, and like his former nest mate he, too, was lucky enough to escape. [Neither crow was ever caught again by Bubo.] Undoubtedly if Bubo had had other crows to sharpen his skills on, Theo and Thor would not have escaped.

One or the other of the crows now always sounds the alarm when Bubo arrives in the clearing, and then they are so alert that a surprise attack becomes virtually impossible. I finally lose all anxiety that they could be taken by him in the daytime. I also do not think they could be taken at night because they sleep hidden away in dense branches of coniferous trees where no large bird could maneuver noiselessly. The crows, now quite experienced, are safe, and Bubo eventually loses interest in them.

The blackfly season is not over, but it is waning. During the next two weeks it will become possible to be outside again without having to worry about their constant attack on every patch of exposed skin. It is also time for Margaret and Stuart to join us for the rest of the summer.

Only Bunny and I are at Kaflunk to greet Margaret coming up the path into the clearing, with Stuart happily bobbing over her shoulder from his perch in the backpack. Bubo comes later in the evening. And hearing our happy reunion inside the cabin, he hoots. Margaret goes outside, and he hoots even more vigorously. Ominous thoughts cross my mind. However, he does not fly down. Maybe he recognizes her.

June 20

Crisis. We are nailed by Bubo while walking down the trail to go shopping in town. He swoops down, lands on the trail in front of us, hoots in anger, and runs directly for Margaret's legs. I hold him off with a pine branch, and then he starts to attack *my* legs.

Margaret is afraid of being attacked when she steps outside in the dark, and wonders if she will have to stay inside a tiny cabin all summer, hostage to an owl. Both of us are even more concerned for Stuart.

When we come back from town Bubo is not at Kaflunk. But the crows are, and they are hungry. I am just about done feeding them, when suddenly they bolt away into the woods in great alarm. Bubo has arrived! He hoots from the top of the large spruce. Margaret and Stuart escape unscathed into Camp Kaflunk.

As I go to get water from the well, I sadly contemplate the grim possibilities. What to do? When I get to the well, about half a kilometer away, he is already waiting there, running at my legs again and hooting angrily. Why attack *me*? I yell at him and beat a stick on the ground, and he flies away and is silent. Now it became clear that he did not attack me *accidentally* when Margaret was very close to me. Does he blame *me* for her "trespassing" onto "our" territory? Does he expect me to keep this hill exclusively for us two?

One way to avoid a fight with him is to be ready and willing to fight back. It works, even if you do not want to fight him, as long as you make believe you do. If he were smart enough to know it was only a bluff, then it would not work. He would bluff right back, and so would you, in turn. The bluffing would escalate into combat, and both combatants would end up doing what both want to avoid. I'm glad he does not understand me better.

Margaret is not confident about being able to outbluff Bubo. Bubo, she feels, has the advantage of weapons and of surprise. Especially in the dark. She is right. However, I am hopeful that she will be able to make peace with him. Stuart, it turns out, holds no interest to Bubo whatsoever. That is a great relief. Apparently children are insignificant to him and not worth bothering about, and women are a threat only if they associate with me.

BUBO's antagonism toward adult human males and his indifference to children are not representative of all owls. In particular, it contrasts with the behavior of a tame great gray owl, "Gray'l," that has been used by Robert W. Nero in Manitoba on numerous fund-raising drives for owl research and owl-habitat conservation. In a recent twenty-month period, Gray'l made forty-six public appearances. This owl (a female?) seems to like people. Nero (1985) writes:

Of great interest is the apparent variable response shown by this owl to different people. Infants and small children, for example, hold her attention far more than do adults, and she tolerates more touching by small people. Once when I knelt down to let a five-year old girl get a close look at the owl I was astonished when the child unexpectedly gave the owl a bearhug and kissed her on the bill . . . the owl never flinched! Particulalry impressive is her response to handicapped people. I watched with incredulity while a blind girl reached out at my suggestion and felt Gray'l from head to toe; again, the owl never budged. On another occasion, a slightly retarded girl practically mauled the owl. Usually, Gray'l ducks away or even shows mild aggression to this sort of handling. Is she, I wonder, picking up cues and tolerating rough handling by such people? For the most part, we now keep Gray'l on a tall perch, and only occasionally do we allow people to touch her, but her variable responses to people is an aspect of her behavior that intrigues me.

And people *like* the owl. Nero continues:

. . . 85% of our time is spent talking to people about Gray'l, owls in general, and other wildlife. All who have been involved in showing Gray'l to the public agree that the most satisfying aspect is watching people's eyes light up when they stand and watch her. And watch her they do, some coming back again and again, some hurrying home and coming back with cameras, some coming back with friends to see the owl.

Nero's acquaintance with great gray owls is long and thorough. He has closely observed hundreds of these birds in the wild, and his book, *The Great Gray Owl: Phantom of the Northern Forest,* is surely a classic.

IN THE afternoon I give Bubo a whole gray squirrel at the log cabin. Instead of feeding on it right in the open by the cabin, as he usually does, he drags it to a secluded spot under a small bush. Even there he does not start to eat until he has looked out and thoroughly checked the sky. Why such interest in the sky? I look, too, and see nothing, but then, a circling speck, a hawk! Bubo gular-flutters in excitement. Is he afraid the hawk might steal his prize? He trusts me and allows me to stay right next to him as he feeds, and then he hides the remainder

Bubo with a gray squirrel,
looking for a place to eat it.

Bubo with a gray squirrel
on the ground, scanning the sky
for hawks. Note the tiny pupils.

near me while I watch him out of the corner of my eyes. If I showed
interest in his cache, he would likely attack me instantly. He seems
to have an impressive ability to read my intentions. Today I have no
intentions of retrieving that squirrel.

June 21

This morning he remains near where he cached the remainder of his
squirrel last night. He hoots at a broad-winged hawk perched in one of
the nearby maple trees. The hawk screams. Why is it hanging around?
Did it see Bubo cache his squirrel yesterday? Bubo flies closer to the
hawk, and after exchanging several screams for hoots, the hawk
swooshes mere centimeters over Bubo's head. That stopped the hooting, but not the resolve. Bubo, now silent, makes a pass at the hawk.
Back and forth the duel continues as the raptors alternately swoop at
each other in silence. Bubo has been sideswiped by hawks before, but
he has never made an attempt to chase back. Undoubtedly the hawk
would steal Bubo's squirrel if it had a chance. But how could Bubo
know this? I am not yet ready to believe that the owl has the awareness to realize that a bird who has never stolen from him before might
do so now. I am surprised that even though he scales his aggression
with humans, depending on whether they are male, female, or infant,
he nevertheless includes predatory birds as suitable objects of attack.

The broad-winged hawk also comes to Kaflunk in the evening. Although it makes at least fifteen passes at Bubo, it cannot get very close
because Bubo crouches out of the way in the thick branches of the
spruce. Here at Kaflunk Bubo does not make a single attempt to chase
the hawk back. Is it because he now has no cached meat here? But
then why does he attack human males here, even when he has no
meat cached?

June 22

Bubo is on the cabin when I get up at dawn, and he begins to hoot
when I come out. He even continues to hoot as I give him pieces of
meat, and he does not tolerate my stroking his feathers. No finger-
nuzzling games today. He is still upset at me, and presumably also at
Margaret (and the hawk?).

One reason Bubo might attack other people is because he is imprinted on me and treats other people as he would any other owl, attacking them to defend his territory. If so, he might not recognize a real owl. I am skeptical, because he did seem to recognize a *hawk* as a threatening creature. Why should he not recognize his own kind?

Tests are in order, but I do not happen to have another tame great horned owl handy to see if Bubo recognizes owls. So I get what I think is the next best thing—a full-length mirror. Bubo looks at himself in the mirror, clacks his bill briefly, and makes a threatening posture. In less than a minute, though, he looks at the top and sides and back of the mirror, and then he ignores it. He was probably not fooled for long. Maybe he is smarter than most songbirds or grouse, who will viciously attack their reflection in a mirror or in a window that reflects like a mirror. We once had a song sparrow who attacked his image in one window of our house every day for about six weeks.

[I later secured a stuffed great horned owl once used by crow hunters. Bubo snapped his bill at it for about two minutes, but after getting no response he totally ignored it. My experiments were inconclusive.]

If Bubo is imprinted on me I am still not sure what that means. As far as I can tell he has not courted me. And he attacks me viciously if I enter the territory where his food is hidden. Would he treat a mate so rudely too? Near a food cache he treats me as he would any intruder, hawk or human, so I cannot tell how he perceives me beyond my being a provider and a threat, depending on the circumstances. He attacks me when I walk to within 6 meters of one of his caches, and strangers are even suspect when they enter within about 100 meters of the cabin where he gets his livelihood. Perhaps with strangers Bubo merely enlarges the boundary that he defends.

He is smart enough not to attack me before I feed him, because then he could not exploit me as his meal ticket. If, indeed, he is imprinted on me and considers me his mate, then the sex roles are reversed. In the great horned owl, it is the male who is the provider, the female who is provided for. If he really is a male and considers me his female, then he has found female liberation remarkably easy to adjust to.

I lure Bubo next to the nest of a pair of white-throated sparrows with young. The nest is well camouflaged in the grass among meadowsweet bushes. There is little danger that he will find it. But what will the adults do with him near the nest? Like the white-throated sparrows of last year, this pair never once attacks him. Throughout the half hour that Bubo is within 5 meters of the nest, both adults perch above him, coming no closer than 3 meters. Their heads turn, but they remain rooted to their perches, continually uttering their monotonous "chinks" like loud, slowly ticking clocks. Bubo pays them no visible attention.

I lead him to a nest of chickadees. He perches directly on the hollow stump with the nest inside it. Both adults call excitedly, but neither swoops at him. Woodpeckers also do not mob him. Is a vigorous mobbing response not necessary because the young are safe inside the hollow trees? Perhaps. But the tree swallows from my birdhouse swoop within centimeters of his head. A pattern would spark my interest. On the other hand, a lack of expected pattern makes it interesting, too, because something unforeseen might be discovered.

June 23

Bubo has been spending much of his time at the log cabin. Every day he perches up on the collar ties under the roof where it is dark and dry, and from where he can survey the clearing and see the woods. The gable ends of the cabin are still not closed—a project for rainy days.

Margaret is anxious to admire our handiwork from last year and to add more to it. But Bubo already starts hooting ominously as we start walking across the field. He is sporting his "mean" look, eyeing Margaret. If looks could kill. As we get closer Bubo flies down and stands guard directly in the door opening, hooting ever louder. He intends to defend that doorstep, and Margaret gets the message. We decide to continue on to the swimming hole instead, and even then he chases us, flying from tree to tree close over our heads. Margaret remains close to me, clutching a pine branch all the way.

In the afternoon I go to the cabin alone to visit Bubo. This time he hoots from the collar ties without coming down to stand guard at the door. He inserts coughing sounds before each series of hoots. In addition, he makes an entirely different sound—one I have never heard before—a series of high-pitched clucks, given after each series of hoots. It sounds like two different birds are in the cabin, one answering the other. I have no idea what the sounds mean. [And since I never heard them again I have no clue to their context.] It is just another of the little puzzles that reminds me that much information has been collected over millions of years and packed into his forty-one pairs of chromosomes (Biederman, Florence, and Lin, 1980; Krishan, Haiden,

Bubo by the doorstep of the log cabin.

and Shoffner, 1965). I suspect the variety of strange sounds reflects a variety of confusing emotions that he is experiencing. He might still be angry at me, jealous, glad to see me, frightened, all or some of these at the same time. I know how to ease his mind from all that confusion. I pull a dead robin out of my pocket. His hoots and clucks change to friendly chuckles. He flies down to me and I sit beside him as he plucks and eats. After he finishes eating he stares directly at me for a full minute, then he comes closer. I hold out my hand a little hesitantly, wondering if I should risk the gesture.

It was worth it. His nibbles on my hand and fingers are rough at first, but they get gentler and eventually become caresses. He continues hooting all the while, even while I'm stroking his head feathers and he has his eyes closed. So he is still disturbed. But we play for fifty minutes, and with time his hootings become softer, until they are almost whispers. (This was the first time he had ever hooted while I was stroking him.) I am glad we have made peace, and I do not break off our play, waiting for him to do so. He eventually does—by pouncing on a roll of twine. He grabs the roll in one foot, and hops all over our piled lumber with the other foot, with some wing assistance here and there. But not enough. He loses his balance and flops on his side. No matter. He is up again quickly, grabbing and releasing the roll. I have not seen him play so exuberantly since he was young.

Having patched things up between us, I am about to leave and step out the doorstep. But he comes, too, and perches all fluffed out in the bright sunshine on the front steps. I recline in the sun in the grass, and he hops down beside me and sprawls out with his feathers fluffed and his wings spread. His head looks huge from the fluffed-out feathers. He stares ahead and remains motionless, as if he were in a dream or trance. After four minutes without moving, his eyes gradually close into slits, and then they close entirely. After another eight minutes, a fly lands first on my ever-present notepad and then on Bubo's bill. He shakes his head, opens his eyes again, and scratches his head with a

Bubo basking in the sun.

talon of his left foot. Looking up he stares at a high-flying jet. Then he hops back up onto the doorstep, looks up over his shoulder once more at the jet, and flies back up onto a collar tie in the shade under the roof. When he begins to preen I go back up the trail to Kaflunk, secure in the knowledge that we still have a meaningful relationship after all.

Bubo is a one-man owl, and Margaret, who once liked him, now prefers the crows. The crows hang out around Kaflunk, and they caw loudly whenever Bubo comes by for a visit. Margaret is glad to have them near, because they are her watchbirds.

The crows follow me into the woods, and we come near a family of blue jays, two adults and some recently fledged young that can barely fly. The adult jays mob the crows in exactly the same way they mobbed the owl—by divebombing from above and behind when their target's head is averted. As soon as a crow begins to preen or to peck at a branch, a jay is certain to attack immediately. But as soon as the crow looks up, the jays stop their attacks, averting those already in progress by veering off in flight. I watch a total of thirty attacks before I, and the crows, leave.

In the evening I go back to the log cabin for a third time, this time with Stuart in my backpack. Bubo is "home" inside the log cabin. I make my introductions, but Stuart is still of no importance to him. I am glad of that, but I am puzzled as to why it is so.

Evening. My other nephew, Chris, and his friend Jeff come up to Kaflunk to visit. They bring blankets and sleeping bags, planning to spend the night. They know *of* Bubo, but they do not yet know much *about* him.

Bubo makes his usual evening appearance at Kaflunk shortly after everyone is settled in and chatting around the table. Unlike other evenings, we hear his hoots from the tall spruce: apparently Bubo knows we have company. He comes down and perches low in the birch for a closer look through the front windows, his head bobbing excitedly. Now he flies to the rear window and presses his head to the windowpane. Jeff approaches Bubo, who is still looking in the window, and he feigns an aggressive gesture. This could have been a mistake. Bubo throws himself against the window with a crash, and Jeff jerks back in surprise. And Bubo is not deterred by a mere closed window; he looks for another way. Round and round the cabin he flies, checking all the windows and the doors for a place to enter. The hooting rises to a crescendo pitch.

Neither Jeff nor Chris seems eager to be the first one out to set up the tent while Bubo stands guard. Can I lure him away with a piece of meat? Nothing doing. He looks right past me into the cabin, as if I, and even the piece of meat, do not exist. It is Jeff he wants. Jeff opens the door, and Bubo instantly lunges from the birch. Jeff trips backward, again surprised, and in his shock neglects to close the door. Bubo barges in, but Jeff recovers quickly enough to grab the broom next to the door. Now Bubo attacks the broom before he is beaten back out the door.

Both boys laugh, but I detect tension in their laughter. Jeff, who weighs 190 pounds, jokes about being "intimidated by a 15-pound owl." (I leave well enough alone and do not tell him that Bubo probably does not weigh a lick over 3½ pounds). Nice bluffing, Bubo! The boys' outward show of humor increases when they realize that they may very well be defeated by this owl. They will not be able to camp near the cabin, nor will it be possible for any of us to sleep inside with the infernal hooting and banging on the windows. It is becoming clear to all that Chris and Jeff must leave, and the sooner they get it over with, the better. It will soon be getting dark, when Bubo will enjoy an even greater advantage than he does now.

For some not-so-strange reason Bunny has already disappeared, so we cannot use him as a hostage again. I suggest that the boys use their blankets as shields or as diversionary targets, like a matador's red blanket. Perhaps they can make it down the trail to the field near the road, well out of his territory, and put up their tent down there. They get ready for a charge. Unfortunately, they lack resolve and hesitate a few seconds too long in front of the screen door. Bubo advances on foot, swaggering up closer to show his annoyance with the situation, and the idea of going out becomes less and less appealing to Jeff and Chris. Finally it becomes downright unpalatable. More nervous laughter. What next?

Bubo is positioned at the door, knowing quite well by now that the enemy must exit that way sooner or later. It appears that it will be later—the enemy's courage has waned and the standoff continues. Bubo's mood is not improving at all.

Necessity is the mother of invention; we must try new tactics. I would not dare to try and hold him, even with leather gloves. But there might be one way to foil him: drop a blanket on him. I go out with a blanket, and he still pays no attention to me: he wants one of those enemies inside, not me. I quickly drop the blanket on him and roll up a very surprised Bubo. In a blanket he sounds more muffled than before, but he is no less animated, possibly even more so. Bulges pop out in all directions.

The boys, not interested in what will happen next, seize the moment by grabbing their things and running out the door and down the trail. Ten minutes later, after they have had plenty of time to make it to the field, I release my grip on the blanket. Bubo unravels himself, pokes his head out without missing a hoot, looks wildly in all directions, and then fixes a long stare down the trail they have just descended. I am surprised he knows they are no longer in the cabin and what route they have taken. He had not *seen* them go out. His look tells me he knows more than I am willing to give him credit for, and I try to hold him at Kaflunk with a bribe. Luckily, he has worked up an appetite and eagerly accepts a large chunk of woodchuck meat. I expect that it will take him at least twenty minutes to tear it apart and eat it, so I jog down the trail to console our company.

The boys are relaxing in the field and about ready to set up their tent, relieved that they made a bloodless escape. But the joy was premature. We hear a hooting. We look up and see Bubo flying down the trail directly toward us. He lands on a large maple, eyeing Jeff without making one blink. It is getting dark, and Bubo's presence is even less reassuring now than it was before, especially after he lands on the ground and starts going after Jeff. Jeff, who is still Bubo's primary target, gathers his wits and wraps himself in a blanket. Right idea, wrong object. Bubo continues his stalking. In the last second, Jeff throws the blanket off himself and onto Bubo. A good maneuver. The muffled hoots sound increasingly strident, and the dancing dervish under the blanket threatens to escape. Quickly Jeff tries to hold him down under the blanket, but somehow Bubo manages to bite through and inflicts a nasty finger pinch. Enough. The boys bolt down the road for a final escape and with a host of tales to tell. I am again left holding the blanket. I resolve to leave a sign on the path: "Beware of Owl."

Until this evening I had thought Bubo's behavior toward Margaret was aggressive. But he had treated her relatively well.

AFTER this episode, Bubo no longer bothered Margaret at all, maybe because he did not realize who his friends were until he met his enemies. He was also tolerant of my daughter Erica, who joined us. He

Sign on the trail to Kaflunk.

was indifferent toward her. The crows, on the other hand, seemed friendly to all but Bubo.

In retrospect, I can think of a number of ways we could have reduced Bubo's being a hazard. For one thing, Margaret would have been safer if she had not run from him and if she had made a point to look straight into his eyes when he threatened to attack. The one thing you *don't* want to do when you are attacked is turn your back and run, as indicated in a recent flurry of letters to the correspondence section of the *New England Journal of Medicine*. The first letter (*N. Engl. J. Med.* 311 [1984]: 1703) came from three Swiss doctors, who noted that jogging is "associated with some harmful side effects, such as musculoskeletal injuries, heart attack, asthma, amenorrhea, frostbite, heat stroke, anemia . . ." and "We have observed an additional danger to joggers: bird attacks." In the last two years these doctors had cared for twelve joggers who had been attacked by birds of prey. "The birds attacked by diving from behind and continued to dive as long as the jogger was in motion. The victims were all men. . . . They suffered from scratches and lacerations up to 13 cm long on the scalp." They concluded that "Joggers should be aware that nature has its own laws and may not allow intrusion without revenge."

A subsequent letter (*N. Engl. J. Med.* 312 [1985]: 1066–1067) by three doctors from the United States offered an alternative explanation to the idea that the avian attacks were an expression of celestial wrath against joggers. One of the three authors was attacked, unprovoked, from behind by a starling (*Sturnus vulgaris*) on the campus of the University of California at San Diego "in broad daylight during November." The victim then withdrew to a safe distance and sat down to witness four other attacks by the same aggressor. These victims were also all male. One was bald. Since it was November the doctors concluded that the bird could not have been seeking hair for nest building, and since it seemed "unlikely that a hairless pate could serve as an effective mirror" (some birds will attack their image in a mirror, mistaking it for a rival), they suggested that "More likely, through faulty aerial reconnaisance [the bird] mistook a hapless monk-head for a large egg in a mobile nest moving in on their territory."

The above explanation, however, did not seem to satisfy a doctor and his son from Berwich, Pennsylvania, who wrote (*N. Engl. J. Med.* 313 [1985]: 330) that they "found no mystery to be solved." The son and his friends had had "similar occurrences while cutting grass." They decided that "perspiration on their heads attracted gnats. The birds, rather than attacking them, were only snacking on the bugs."

Without commenting on the medical men's theories on bird behavior, insect biologist Thomas Eisner of Cornell University at least offered a cure for the joggers' affliction in the fourth (and perhaps the last) letter (*N. Engl. J. Med.* 313 [1985]: 1232–1233) on that topic. Eisner relates having suffered through attacks of Australian magpies (*Gymnorhina* spp.) during their nesting season in Canberra. To protect himself he took his cue from the British biologist A. D. Blest (1957), who showed that many birds were frightened by the eyespot patterns that decorate the hind wings of many large moths (Schlenoff, 1985). In addition, predators seek the advantage of surprise, so if *they* see eyes it is likely that the eyes see them, too. Therefore they have an inhibition against attacking as long as an eye is visible. Some moths, and many other insects, have probably evolved fake eye spots as protection from birds. Eisner wrote, "What works for insects, I thought, might work for me, and . . . I decided to affix two fake eyes to the back of my cap." Figure 1 in his publication shows two dark eye spots, with prominent white margins, at the back of a jogger's cap. The intent is that these fake eyes will reduce one of the many hazards that joggers face: birds on the attack will presumably be startled and hesitate before an attack. All joggers will be grateful.

Eye spots on the underwings of a saturniid moth, here displayed during the startle response.

The biological invention of these moths, which Blest discovered by pursuing a typically "esoteric" question just for the fun of it, may have a totally unanticipated and much more important application than protecting the pate of a few joggers: it protects airplanes as well. In 1985 alone, 406 birds smashed into planes belonging to All Nippon Airways, causing half a million dollars worth of damage. So Nippon gave some of its planes the evil eye by painting eyeballs on the conical fan spinners of the jet engines. The frequency of air strikes by birds went down dramatically, and Nippon has now patented the idea.

What works for insects—and now perhaps joggers and airplanes—has also worked for some owls. The dark eye spots with prominent white margins are found not only on the underwings of some moths but also on the back of the heads of the pygmy and ferruginous owls. Both of these small owls are active in daylight, and both catch birds. But nobody has yet determined, as far as I know, whether or not the fake eye spots of these owls give protection from mobbing, or whether they confuse their prey and increase their success in catching small birds.

Matching Wits

Both crows, and Bubo, follow me through the woods today. One might think: what a motley crew—one man, one owl, and two crows! But we proceed smoothly. The crows now scold Bubo only briefly on first meeting him, and afterward they pay little attention to him. Bubo no longer makes any attempts to capture them. He usually follows me unobtrusively at a distance, and I sometimes don't know if he is following at all, until he appears from nowhere to fly up ahead. The crows, on the other hand, wend their way through the forest close to me, sometimes following at my heels like puppies, sometimes flying up to my shoulder to steal a ride. I like them there. I love their whispering, cooing noises in my ear, and they catch almost every deer fly that ventures to circle my head.

The crows can be fast, agile, and powerful flyers who negotiate through the branches like large unleashed black arrows, adroitly using their tails as rudders to steer. The owl, on the other hand, picks his way carefully and clumsily. The crows launch themselves with little or no conern for the next perch, and the path to it. But the owl seems to contemplate every move, as if deciding whether or not he can get there from here. The crows know they can. The owl must weigh whether or not he can negotiate this or that turn, get enough lift to fly up to that branch, or slip between those branches without hitting his wings.

We flush some young white-throated sparrows from a nest on the ground. Bubo notices a young bird trying to escape, and catches it and swallows it quickly. The adults' protests are of no avail. He pays no attention to them. The parents' protests and mobbing might protect their young before they are discovered, but not afterward.

We come near a nest of hermit thrushes with young. The parents ignore me, but continuously mob Bubo the full half hour that we remain there. They divebomb him fifty-three times during that time. The thrushes snap their bills as they zoom by the back of his head, making him turn. By watching one bird closely he can prevent that one from attacking, but thereby he leaves an opening for the other, and between the two he is distracted, and his eyes do not search the ground where the young are that he *can* catch.

In the half hour that the hermit thrushes mob Bubo, only one other bird was attracted by the loud commotion they made—a red-eyed vireo. The vireo looked at Bubo, called just once, and then left. [Indeed, I did not see a single flock of birds gather to scold Bubo even once in three summers. They only scolded if he came near their young, whether they were in or already out of the nest.]

I find a wood frog and call Bubo, who immediately comes flying through the forest to land by my side. I release the frog in front of him. He walks toward it. The frog responds by making four hops in quick succession in seemingly random directions. The first hop is away from him, and Bubo starts after it; but in the meantime the frog has already reversed direction and has jumped right and left as well. It is now crouching down, blending in with the dead leaves. Bubo is baffled. He has no idea where the frog ended up, even though it is only within a half meter of him, and so I grab the frog and release it a second time. This time Bubo pounces right on target, immediately crunches the head, and eats it all after tearing it into little pieces. Do wood frogs really taste that much better than the bullfrogs he disdained last summer?

A little later I find wood frog number 2. I call Bubo again. He catches this one, too, and it slides down twice as fast as the first, with less dissection.

On another occasion, some days later, I found still a third wood frog. This one played possum after I caught it, although only when it was camouflaged by being belly-side down. When I turned the "dead" frog belly-side up, it immediately righted itself and crouched down with its legs tucked in closely. I picked it up again and called Bubo, tossing it onto the leaves. He landed on it, and it still did not move a muscle. After several seconds Bubo let it go, acting as if he had made the mistake of pouncing on a leaf. He looked all around, oblivious to this frog who had, on this occasion and possibly on others, done the right thing for itself. One way to succeed in evolution (as in business) is to follow closely the already successful—the tried and the true. The other is to be entirely different—to be unique and innovative. This can be risky, but it has its rewards, as it had for this particular frog. I do not know which is the most common defensive strategy of these frogs; but it is significant that, like the bullfrogs Bubo came in contact with earlier, they do not all act alike.

There is no absolute best attack behavior for the owl, because there is no absolute best escape behavior for the prey. Frogs have a varied strategy. If there were total predictability in their responses, then it would be easy for the predator to use the appropriate behavior to make unfailing captures. Samuel Butler said, "Logic and consistency are luxuries for the Gods and the lower animals," but consistency in escape behavior is a luxury they cannot afford.

JUNE 28

Since dawn the crows have been singing their curious warbling songs, but they caw only when Bubo arrives, as he does almost every morning.

He could be catching some of his own food now because he is not always hungry any more. But if he acts sensibly and in his own interests, he will likely mooch off me as long as I give him an easy meal. His mooching is more a sign of his good sense than an indication of his ineptness in prey catching. But he probably does do some hunting even when he is not hungry.

BENT (1961) reports observations of O. E. Niles who found "several full-grown Norway rats with their skulls opened and their brains removed" in a nest of great horned owls. Niles also noticed the bodies of many rats on the ground around the tree, and "out of curiosity counted them and found the bodies of one hundred and thirteen rats, most of them full grown. They all appeared to simply have had their skulls opened and the brains removed; and from their undecayed appearance, must all have been captured within the previous week or ten days." Perhaps those owls killed for the fun of it, and/or their tastes were highly refined because of a superabundance of food.

Although it may seem that the owl's, or any predator's, hunting could seriously deplete the fauna to nature's detriment, it is more likely that it will enhance the quality of nature as we generally per-

ceive it. One of nature's major appeals is its endless variety. Of the innumerable species in any one place, each has the inherent tendency to
over-reproduce until it degrades the environment and crowds out the
others. Predators key in on the most common prey. They learn how,
when, and where to catch it. They do not generally attack the rare and
unfamiliar when enough of their regular prey is available. In a sense,
then, the predators are the equalizers who may give the underdogs,
who might otherwise be crowded out, a chance to exist. Nevertheless,
exotic introduced predators have caused innumerable extinctions.

July 15

Unlike the owl who spends most of the day and night fluffed out and
at ease perching in his birch, the crows seem to delight in flight. They
are good at it. They chase each other, and also fly solo, careening
around the clearing and through the tops of the trees in the woods in
their dazzling aerial play. In contrast, Bubo's play has always been
with his feet, in pouncing and grasping, but never in flight. Undoubtedly, both species of birds are practicing the skills that are or will be
important to them.

Bubo should be impressed by the crows' virtuosity in the air and by
their quick alertness. Maybe he is. He has ceased even to follow them
with his eyes, much less with intent to catch them. The crows can
sidestep him as easily as I can step off the road when a car comes, and
because they always stay close to each other they are doubly safe: if
one does not see him, then the other one will. The first one to see him
sounds the alarm and after a bout of mobbing, all obligations have
been met, and each is then on its own to remember or keep track of
Bubo.

The crows seem to be more quick-witted than the owl. They act as
though they *understand*, and for all I know perhaps they really do.
They show a behavior that is suspiciously logical. Given dry bread or
crackers, they try to break it apart, and if it does not crumble easily
they may fly to the nearest water and dunk it, softening it before eating it. They do not dunk bread and crackers that they hoard.

They also show a crude kind of quantitative reasoning or reckoning.
Large, whole saltine crackers that must be dunked or picked apart
take time to prepare for eating. Crumbs do not. When I give them a
mixed pile of many crumbs and whole crackers, they do not grab the

A crow is brave when the owl has his head turned (top),
but it takes flight when noticed.

biggest piece of food first, as Bubo invariably would. No, they opt for the crumbs that can be eaten quickly. But if there are only a *few* crumbs and only one cracker, they hasten to beat each other to get that cracker, rather than pick on some crumbs. Perhaps they realize that they can handle their cracker at leisure somewhere else and eat it all by themselves, but if one stayed to eat crumbs it would end up with less than the one who took the cracker. Somehow they are able to evaluate the odds at a glance.

July 20

A run-over raccoon provides me with a temporary surplus of meat that will spoil long before Bubo and the crows can eat it all. I do not like to waste meat, and my failure at preparing edible clams has robbed me of confidence in cooking coon. Perhaps I can use the meat instead for an experiment to learn if Bubo will cache all his surplus in the same place. Will he remember where he put it all? The crows, who are sure to get in on the act, might give added fun.

As I leave Kaflunk at 7:30 A.M. with a plastic bag full of chunks of meat, the crows are, as usual, alternately or simultaneously underfoot and on top of me. For their efforts, and my peace, they get a few chunks of meat each to fill them up.

Bubo has been absent for several days. I am glad that he is becoming independent. But I am also glad to see him back today, perched on the wood-chopping block by the doorstep, where he is making his low, contented grunts. I give him a choice of several chunks of meat. He takes the largest, and he takes his time tearing off and eating only tiny pieces of it.

The crows, even after they are full, will not leave. Thor flies off with a chunk and caches it 30 meters away. In less than a minute he is back. This won't do. I stop trying to satisfy the crows, who cannot be satisfied. Realizing quite quickly that I won't give them more, they now both approach Bubo with the obvious intent of stealing *his* portion. They dare to approach him within 30 centimeters, and from that distance they make short lunges at him while cawing loudly. Bubo hisses, and they soon leave him alone, but not before Thor gives him a hard yank on his tail.

Now I prepare a pile of chopped-up meat for the crows, more than they can possibly consume in one day. Both crows converge on the pile and hastily stuff their bills, and with bulging throats they fly off instantly to cache the bonanza. They fly back and forth, taking load after load without a moment's hesitation before departing. Each load is cached in a different place. They seem to have resolve and conviction and know where to go each time, but I do not know if it comes from knowledge of all the best caching places or from a lack of differentiation of the possible alternatives. Working at a frenzied pace, as if to see which one could lug off the most, they take a few seconds at each depot to cover their booty with a leaf or two, or moss, or anything else loose and handy. Still, they do not trust each other; when one crow walks near another's cache, the owner comes flying to retrieve his or her prize and hide it elsewhere.

Bubo, meanwhile, continues to pick at his meat. I sit beside him and soon become irritated by the crows who have finished their hoarding and who now try to steal my pen, tear my notepad, and dip their bloody bills into my coffee. They act as though Bubo no longer exists, and he completely ignores them as well.

Bubo all of a sudden stares into the woods in the direction of Ka-flunk. He fluffs out his feathers and clacks his bill in a full-blown threat posture. Then he flies off, leaving his piece of meat, which he could have easily carried. The source of his irritation is more than 20 meters away, walking softly and meowing weakly. The crows usually caw loudly at Bunny, but not when he is this far away, where he is no threat. They are unconcerned now, and take the meat Bubo has left in his hasty retreat.

I pull another large chunk of meat out of the bag and call Bubo. He comes as I had anticipated and flies off with it, landing on the stub of a broken-off fir tree. And there he perches on top of his meat. Where shall he cache his prize? He makes soft grunts of contentment and scans the ground in all directions, peers into all of the surrounding trees, and looks up at the sky. He again scans the ground. Decisions, decisions. Unlike the crows, he is not a master of quick decisions. Bubo, faced with many simultaneous opportunities, is paralyzed by inaction. Ten minutes go by. Twenty. Thirty. Still no action. But his gular flutter on this cool overcast morning indicates that he is excited; he is certainly not gular fluttering due to overheating from exercise. Perhaps he just leaves nothing to chance.

Meanwhile, I watch a little wasp with blue-black wings dragging an immobile black spider over the tops of the ledges. It drops the spider, dashes ahead on spindly legs to a crack in the rocks, comes back to the spider, drags it some more, drops it, dashes back to the crack, and then drags the spider all the way in.

Forty minutes. Bubo is still looking all around. No decision yet. He takes his time, as if heeding Francis Bacon's dictum, "Nature is a labyrinth in which the haste you move with will make you lose your way." Flies are now coming to his meat, such is his lack of haste. But he only shakes his head in irritation when they come near his head.

Forty-six minutes. A solution has been found. He flies with his meat trailing from his talons. I run after him in time to see him land on the ground far ahead, and hopping on his left foot he drags the meat by the right and caches it under a little spruce.

As Bubo is dragging his meat, Thor is already on the ground observing him from nearby. Bubo is unconcerned. Unlike the crows, he does not cover his meat. He merely shoves it into a secluded spot, and leaves it at that. But it is not left for more than a minute before Thor goes walking over to it. Bubo chases him away once, but then comes to me to receive his second piece of meat. Thor, of course, uses the opportunity to retrieve the piece he has just left.

He was slow to cache his first meat, but not because he does not care for more. He takes not only a second piece, but also a third—and a seventh. He hides each of them in a different place, but they are all within 15 meters of each other. Although it took him forty-six minutes before he finally hid the first piece, subsequent pieces are hidden in three, two, one, seven, two, and two minutes, respectively. As far as I know, the crows, who stay close by, find none of the other caches. They have by now become disinterested in Bubo and his activities, preferring to tear loose layers of moss and lichens from the rocks, possibly finding insects.

After each time that Bubo caches a piece of meat, he returns to a dusty patch of dry ground near the center of his caches. Here he alternately fluffs out his feathers and takes a dust bath, or attacks nearby moss hummocks with his talons. Wanting to keep him company, I sit beside him near the center of all of his caches. He must know I have no intentions of robbing his caches, because he ignores me. He reveled in the dust for bouts of a few seconds. The rest of the time he stands still, patroling with his eyes in all directions. He continues to gular flutter, showing his excitement. Meanwhile, Theo has returned to Ka-flunk, and Thor sits on a red maple branch no more than 8 meters from Bubo. Thor turns his back to him and closes his eyes, occasionally opening them when Theo calls from the cabin.

After half an hour Bubo still has not moved from his observation post on the ground near his meat caches. Does he remember where all of his caches are? I stand up and take three steps toward one. He stands up tall and sleeks his feathers. He knows my intent immediately. He hoots, and I do not mistake his message, as he comes flying directly at me. I dash off, losing him in the woods on my way back to Kaflunk. Yes, he remembered at least where *this* cache was. I do not care to test all the others.

Evening. Bubo is perched in a spruce near his caches, and he flies at me as I merely approach the general area. I leave.

WATCHING an owl all day long might seem to be a lonely task. Indeed, I would have been thrilled to have someone close to me beside me, someone who shared the interests and the sense of excitement and discovery. That would have made my task even more fun, and it would have given it even more meaning and satisfaction. Sharing a passionate interest makes it grow. But a passionate interest that may be admired from a safe distance is often thought by those who are closest to you to be a self-centered indulgence. Worse, it is difficult for others to criticize the passion directly, and so the resentment rears its head at unexpected moments.

JULY 21

Bubo joins us at the cabin in the early morning, and the crows mob him viciously for about an hour, then totally ignore him for the rest of the day. Strange; recently they have usually mobbed him for only a few minutes each day.

I had expected that Bubo would stay to guard his meat, and so I am surprised to see him at Kaflunk. I offer him a meadow vole and a chipmunk. He swallows both. That is even more surprising, given the surfeit of food he had yesterday.

Will he defend his caches of yesterday? I (and the crows) go to the cache area. Strangely, he does not rush after any of us, and he comes only after I call him. I walk directly toward the same cache where he attacked me yesterday. Bubo looks on nonchalantly and makes little grunting sounds. It seems he knows something I don't. Indeed, the meat is gone. I make the rounds to all of the caches, and he continues to watch me and make chuckling grunts. All of the caches are empty. He could not possibly have eaten all that meat. Perhaps he moved it. Or did the bear that has lately been foraging in the nearby raspberry patch take it?

July 22

I set Stuart down beside me in front of a thickly laden raspberry bush. He is gurgling happily while we pick our daily ration of berries, and he watches Bubo on the birch next to us. The crows are cawing a half mile away at another clearing. I wonder how much this little boy, not yet one year old, is taking in. His windows to the world are just opening, and it is exciting to think that I am here to help him find them, and not just berries.

·ALTHOUGH Bubo did not appear to be impressed by Stuart, the reverse was true. A year and half later, after not having seen Bubo for over a year, he still showed great enthusiasm for owls. Field guides of birds became his favorite books, and he would ask me again and again, as we flipped the pages, to go back to show him the "auuls." He also became interested in other birds, and he could identify parrots, hawks, ducks, gulls, and others in the books. By the time he was a little over two years old he could identify chickadees, robins, grosbeaks, blue jays, and crows on our frequent walks in the woods. Apparently one's interests and windows to the world can be developed quite early in life.

July 23

Bubo comes and perches on my leg as we rest on the ledges. Theo is cawing loudly directly in front of him, while Thor stalks up behind him and yanks on his tail. Bubo wheels around and hisses. In less than a minute, Thor does it again. Now he makes no more pretext of stealth, but waddles up with his head held low, feathers erect, making his deep gutteral caws. Bubo stares at him with mad marigold eyes, and Thor stops advancing.

JULY 26

Jogging down a country road so small that the people who live nearby actually wave at you, I pass a field where a group of crows are cawing. Thor had not been near the cabin for two days. Without breaking stride I yell, half in jest, "Come here, crow!" and to my great surprise one of the crows leaves the group, flies after me and above my shoulder, and then lands on my outstretched hand. It is Thor! The other birds circle twice and caw loudly, perhaps awed by this brazen bird they had adopted. (The people who are watching are awed, too.) Thor follows me home, flying closely behind. He is not hungry. Apparently he is quite capable of fending for himself.

LATER on in the summer the crows stayed more and more with other crows, often bringing their wild flockmates to us. However, the longer they stayed with other crows and heard them scold us, the more they became shy of us. Perhaps they were now learning from *them*, believing what they were told, rather than what they themselves had experienced. There might be a disadvantage in being a social animal and learning from others. But there is also a great advantage—crows who have never experienced being caught by a great horned owl will learn from those who mob owls to fear them just as my tame crows were now being successfully taught to become shy of me.

I was surprised to see that wild crows readily tolerated the two I had raised. There was no prejudice. Perhaps crows have no idea of how others "ought" to be. On top of that, they are beautiful. I like crows. I like them a lot.

The Tail Molt and Evolution

BUBO has only recently regrown his tail. For most of the past month he was first without one, and then only with a very short one. One week in June several tail feathers in the center of his tail had fallen out, and within a week all twelve of his tail feathers followed suit. I did not think it of note at the time, but his clumsiness in flight over the last month got me to thinking about the tail molt. Hawks like the goshawk, Cooper's hawk, and sharp-shinned hawk have short wings and long tails. These hawks are the antithesis of *Bubo* in their hunting strategy. They pursue agile prey—other birds—at great speed through the forest, and they remind you of an arrow with a homing device. They ultimately make successful captures because they can stay close in pursuit of fleeing prey by executing sharp and sudden turns. For this their long tail is vitally important as a rudder. They could probably not get along without it. Why has an owl evolved the ability to live at least temporarily without a tail?

Birds have evolved rather specific molt patterns (Stresemann, 1967), and a simultaneous tail molt speaks about evolution. It tells us something about the hunting behavior of owls over millions of years. Ernst and Gretel Mayr (1954), who first described the tail molt of owls by studying museum specimens, wrote "A gradual tail molt, involving successively one feather after the other, is presumably based on a more complex physiological mechanism than a simultaneous tail molt . . . the complex mechanism can stay perfect only if maintained by selection. If such selection should relax, a simultaneous tail molt might take its place."

During Bubo's third summer his tail feathers also molted in a short period of time, but the molt was not as cleanly synchronous. In the first year he shed his middle tail feathers last, several days after the others. In the second year, in contrast, the feathers in the middle of the tail fell out first. On June 21 only one tail feather was missing out of

Bubo visits Kaflunk. Note absence of tail, due to molt.

the middle of his tail. By three weeks later, all of his tail feathers had been shed, and all of the new feathers had appeared. In another two weeks the four middle tail feathers were almost normal length and the rest were about 5 centimeters back, in a "step." Just prior to the molt he was unusually hungry. On June 21 and 22, for example, he ate one chipmunk, three red squirrels, one long-tailed weasel, one mole, two catbirds, one cedar waxwing, one mouse, one robin, and one Canada warbler. (I had been in training for an ultra-marathon, and at this point I was able to collect road kills from over a 20-mile stretch of road on most days.) Perhaps Bubo's hunger was related to the large metabolic demands of growing new feathers so quickly. I wondered if the molt would be more prolonged if he had not been fed so well.

Apparently the tail molt is not an invariant characteristic. There is a lot of plasticity of response. Katherine McKeever at the Owl Rehabilitation Research Foundation in Ontario, Canada, wrote me: "There is no doubt at all that stress—whether metabolic (starvation, poison) or psychological (continuous anxiety in unsuitable captivity with irresponsible and frequent handling) can utterly defeat the normal moult pattern—even preventing it altogether." Bubo's molting, however, could hardly have been the result of abnormal stress.

I consulted Robert Nero in Winnipeg, Manitoba, for additional information on the tail molt of owls because of his extensive experience with owls, and he wrote me the following:

> Tail molt in owls varies with species. With the Great Gray Owl there is a lot of variation. We even had one adult female shed her entire set of retrices while she was still laying eggs and incubating— that is against the rules! . . . We have seldom found two wild great horned owls with similar patterns of new and old feathers. . . . Many wild-captured great horned owls have all new tail feathers (winter period). And an adult male I skinned this week (Dec. 19) had a lot of pin feathers although it was obviously not getting enough food. . . . In some cases old feathers that are retained are so badly worn they can hardly serve any purpose. Doesn't make sense, does it?

Dr. Nero added that he doubts that synchronous tail molt is a regular occurrence. But since it varies so strongly with individuals, it is also clear that a precise pattern is not of extraordinary value to the birds, or else a more precise pattern should have evolved. In short, I conclude that there has not been a great penalty to owls throughout evolution to lose the complex physiological mechanisms necessary for a gradual tail molt. And that provides a vital link to my ideas on mobbing behavior (see Appendix 2).

The End of Summer

JULY 28

The hint of crimson on some of the red maple trees in the swamps is a breath of fall. Already the purple fireweed in the old cellar hole near the log cabin is producing airborne seeds, and the field is becoming cloaked in the bright yellow bloom of goldenrod.

The field now supports a large crop of grasshoppers. Not one will live to see the winter. Green katydids sing their love songs from the still-green marsh grass, and field crickets call every evening from underneath logs near the cabin. And from the woods on warm days one hears the buzz-saw drone of the cicadas. All these insects are occupied with reproduction, leaving eggs that will start the cycle anew next year.

Yesterday I saw a potter wasp insert a slender brown caterpillar into her delicately symmetrical urn of clay with flanged opening that she had constructed under a flap of birch bark on a log on our woodpile. Today the opening of the jug is sealed. Preparations for winter are going on all around, and the summer has already receded so much into the past that I savor the present even more.

In contrast to the insects, the birds are strangely silent at this time of year. Most have raised their broods, and they are now preoccupied with foraging and getting fat for the southward migration. Small flocks of them are already gathering.

AUGUST 8

Our log cabin is nearly built. The windows are in, and the chinking can begin next year. My training went well. [Somewhat later I set the new U.S. national record at 100 miles.]

It is a beautiful, warm, clear day. We celebrate finishing our work by going down to the swimming hole. The sunlight filters through the

light green leaves of the maple canopy, and they seem almost luminescent when you look through them from below. Down at the
stream the dragonflies are hunting small insects above the calm water
of the swimming hole. Water striders dimple the surface, leaving tear-
drop shadow marks on the sandy bottom. Small minnows swim in the
shallows, and they nibble at your toes as you wade in the cool fresh
water.

The crows have followed us today at a distance. Now they are exploring up and down the brook, feeding on insects along the edges.
Only Bubo is missing. We have not seen him for several days.

Then from out of the forest comes the familiar, resonant "who-
whoo—" Is it Bubo or some other owl?

A few minutes later, a big owl lands on the white birch by the side
of the pool. It *is* Bubo! He peers down on us, and the crows stop their
stream exploration and mob him immediately. As usual, after having
given him brief but concentrated attention, they ignore him completely. They hop from rock to rock in the brook, and Thor lands in a
shallow pool and splashes until the spray flies. Bubo continues to peer
down from his perch. He hoots some more, and then he also hops
down into the water and thrashes his wings, adding his own vigorous
splashing to that of the stream flowing over the rocks. Then he fluffs
himself out on one of the larger rocks in the sunshine and preens himself. Margaret and Stuart are within several meters of him but he pays
no attention to them.

Bubo flies upstream, and both crows immediately give chase, although they had paid no obvious attention to him for the hour or more
that we have been splashing in the pool.

Well refreshed, we travel back up the trail to Kaflunk. Bubo does not
follow. We hear what might be a muffled hoot in the distance when
we arrive at Kaflunk. I call him repeatedly, but he does not come.

POST SCRIPT

I came back to Maine two weeks after we had returned to Vermont, to
see how Bubo was doing. He came to greet me at the cabin after I
called. Fearing he might not catch enough prey after the freeze-up, I
captured him and brought him back to Vermont, where he spent the
winter in a new huge outdoor aviary in my backyard.

Bubo, being friendly and curious, in the woods.

In the spring, after the snow had melted, I brought him back to Ka-flunk. Later, during the summer, I was at the camp only sporadically, without Margaret and Stuart. Bubo often came to see me when I did come, but he did not beg and he often ate little of what I did offer. The crows were now totally on their own. One of them came by occasionally and bowed its head and made cooing sounds from a maple tree directly above me. But it would not come down to be fed. I felt good to know at least one of the pair survived the winter.

During Bubo's third summer, like his first, he did not attack any of the visitors who came near the cabin. But he was as aggressive as ever toward me when I came near food I had given him and that he had cached. Maybe he was less aggressive near the cabin because the immediate surroundings were no longer the sole source of his livelihood: it was not so important to him, and so he defended it less strongly. Since he could not know that other territories might not be any better than this one, I feared that his high expectations resulting from my great solicitude toward him might now make him dissatisfied with the new circumstances, and he would leave. Indeed, by the end of August he failed to show up at all. But that was as it should be.

That summer was painful to me not only because it was the end of my relationship with Bubo, but also because it was a summer without Margaret and Stuart. Margaret had already made it clear at the end of the previous summer that she would not be returning to Kaflunk with me. Fate had brought her into my life, as it had Bubo. The nest that the nestling Bubo was destined to fly from had been shattered that winter because of a series of natural events: a frost had made the sappy pine limbs brittle because a tropical storm veered north at the "wrong" time and encountered cold air; Bubo's parents had chosen to nest on an ill-fated limb; and finally, one snowflake too many had fallen—gently, innocently—on a branch. The last snowflake was not at fault. And so it was with Margaret and me. Sometimes the reasons for an occurrence are so complex and out of our control that it helps to think of them as fate, and all of us creatures of the Earth must live with fate and embrace it.

APPENDIX I: BOOKS AND BACKGROUND

ALTHOUGH I had intended to provide an appendix on the general biology of owls to serve as a context for my observations on Bubo, a perusal of the literature quickly convinced me that there is little need for even more repetition on what is already amply provided elsewhere. For the last twenty years, on the average, one book on owls per year has been published in the English language, suggesting that the general biology of owls may already have received more exposure than that of any other bird. Therefore I will provide here only a short guide to the recent books on owls, for the reader who would like more information about these birds than I have provided. This will be followed by a brief overview of the great horned owl in particular.

For general background on owl biology, I recommend *Owls, Their Natural and Unnatural History* by John Sparks and Tony Soper; *Owls of the World, Their Evolution, Structure and Ecology* by John Burton; and *A Natural History of Owls* by Michael Everett. The book by Burton includes beautiful color photographs of many owls from all over the world.

The characteristics and habits of the eighteen North American species, as well as a large potpourri of owl facts, are given in *Owls* by Tony Angell; *The Owls of North America* by Karl Karalus and Allan Eckert; *The Book of Owls* by Lewis Walker; and *Owls by Day and Night* by Hamilton Tyler and Don Phillips. Heimo Mikkola provides a considerable wealth of research details on owl ecology in *Owls of Europe*; David Fleay, in *Nightwatchmen of Bush and Plain*, gives ample information on Australian owls.

Three other excellent books deal with individual species. Ronald Austing and John Holt, Jr.'s *The World of the Great Horned Owl* gives a general account of *Bubo virginianus* in the wild, enhanced by many personal anecdotes and observations and black-and-white photographs. *The Barn Owl* by D. Bunn, A. Warburton, and R. Wilson summarizes the authors' combined thirty-eight man-years of barn-owl observations in northwest England. Finally, the finely written book by Robert Nero, *The Great Gray Owl: Phantom of the Northern Forest*, gives a vivid, personal account of research and observations of these birds in the wild. The book is illustrated with stunning color and black-and-white photographs of wild great gray owls by Canadian nature photographer Robert R. Taylor.

Besides the many highly readable scientific or scientifically based books, there are also some with a primarily literary orientation. *Owl*, by William Service, the little book about a screech owl "the size of a beer can and the personality of a bank president" who resided in the author's study, is an especially delightful read. Jonathan Maslow's enthusiasm for watching owls in the wild is palpable from his book *The Owl Papers*. Peter Parnall's artistry shows in his book with A. Cameron, *The Nightwatchers*. Finally, among the children's books on owls is Farley Mowat's *Owls in the Family*, a story about his two pet great horned owls named Wol and Weeps.

Owls may be secretive, but they have attracted more than their share of scrutiny. In 1978 Richard J. Clark, Dwight G. Smith, and Leon H. Kelso, in a monumental effort, compiled a comprehensive guide to the scientific literature relating to the owls of the world. They cite 6,590 research works on owls, including 1,064 on the genus *Bubo*.

THE GREAT HORNED OWL

In their book on American owls, Tyler and Phillips (1978) write: "The deep, resonant call, a *hoo hoohoo hoo hoo*, of the great horned owl is probably the most familiar of all bird songs. It is associated with harvest moons and chilling frosts of autumn, and as the year grows colder and December turns to January hooting increases as males and females speak to each other; the female's is a shorter and higher sequence of notes with a less regular cadence."

The great horned owl is not only one of the more common owls, it is also the most widely distributed. It occurs throughout the United States and Canada and in the timbered regions of North, Central, and South America; its range spans from the timber line in arctic Canada and Alaska to the straits of Magellan in southern Chile. Not all great horned owls live in forests. In the deserts, for example, they find shelter in cliffs and in rock crevices.

Throughout its distribution, the species is subdivided into various races, or subspecies. The 1931 American Ornithological Union check-list recognized ten subspecies in North America alone. The various subspecies differ somewhat in coloration and size. The northern ones are the largest. That from the Arctic, *B. v. subarcticus*, is very light, and the one from the southwestern deserts, *B. v. pallescens*, is also pale. In contrast, the subspecies from the Pacific coastal forests, *B. v. satyratus*, is quite dark. Other subspecies occur throughout South America (except in the Amazon Basin), all the way south to Tierra del Fuego (Traylor, 1958). The unmistakable field marks of all great horned owls is their large size (about that of the red-tailed hawk), prominent ear tufts, and white throat patch. No other North American owl has all three of these field marks simultaneously. Its general size statistics are: length, 41–58 cm; wingspan, 89–140 cm; weight, 1.4–2.3 kg.

Owls and other predatory birds have "reversed" sexual dimorphism: females are usually larger than males. Several hypotheses have been suggested to account for this difference (Andersson and Norberg, 1981). One theory is that sex-role partitioning is advantageous where one parent guards the nest while the other forages. The female, who incubates the eggs, also guards the nest, and she is therefore favored by large size. The male may be smaller for two reasons: because it makes him more agile in the pursuit of prey; and because after the chicks leave the nest and the female ends her guard, her larger size will enable her to hunt prey of sizes other than that depleted locally by the male. It is not known just how much, and for how long, a pair of owls the size of great horned owls deplete the local fauna.

The snowy and great gray owls may be larger than the great horned owl, but they are not as powerful. Tyler and Phillips (1978), in their comparative study of American owls, emphasize the fierceness and strength of the latter: "There is a terrible power in [its] beak and talons. . . . This owl can rip the wings from a marsh hawk taken in mid-air as easily as some other owls remove moth wings from their prey. . . . The spotted owl of the West is so thoroughly afraid of this predator that one of the requirements of its own habitation is that there are no great horned owls anywhere nearby. . . . [It is] a top predator who will take a skunk or tomcat as readily as a mouse."

The "typical great horned owl" *B. v. virginianus*, Bubo's subspecies, is found in eastern North America from Ontario and New Brunswick south to Florida and the Gulf Coast, and west to Wisconsin and eastern Texas. This bird is particularly well suited to heavily forested regions, and it is a relatively common bird in woods stocked with small game.

In the East the owl most commonly nests in the expropriated nest of a red-tailed hawk in some tall white pine, although it will also nest in hollow trees, if available. In Florida, it commonly nests in bald eagle nests, while in Texas and the West it commonly nests in a rocky cave or a ledge. In the prairie regions and in Colorado it sometimes even nests on the ground in a simple scrape.

The great horned owl is the earliest-nesting raptor. In New England the female lays her eggs in February, and sometimes as early as January. Incubation requires about twenty-eight days, and is done almost always, if not entirely, by the female. Clutch size varies from one to five, though it is usually two in the northeast. January and February are usually the coldest months of the year in New England, and the incubating female is commonly surrounded by snow. There have been reports of eggs freezing in winter, or being flooded in early spring (Bent, 1961), in which case a second clutch is laid, and sometimes a third if the second is also destroyed.

The hatched young have a sparse, pure-white down, which is replaced in a week or so by thicker gray down. The new down retains the old at the feather tips, which eventually wears off. The young remain in the nest for six to eight weeks, although they are unable to fly until they are nearly three months old.

The adult great horned owl is a "top" carnivore and has few predators. By most accounts cottontails and hares are its favorite prey, but the birds are opportunists and in some parts of their range subsist on mice, woodrats, and birds. About half the young do not survive their first year, probably due to starvation. In captivity, however, Bent (1961) reports that a bird has lived for twenty-nine years. As in most other birds, the owl's population is largely determined by food supply and/or the habitat that supports the food supply.

APPENDIX 2: ALARM RESPONSE TO PREDATORS AND MOBBING

IT HAS long been known (Marler, 1957) that when an owl is found by any of a large number of other birds, the discoverer gives alarm calls, and soon a vociferous "mob" of birds surrounds the owl. The mobbers appear to put themselves at risk, and one wonders why they do not remain quiet and leave as soon as possible so that they are not detected. This is a question of some interest to evolutionary ecologists. In order to answer it, it is necessary to ask some other questions: Do the mobbers really put themselves at risk? And who benefits from mobbing, and why?

There is no doubt that to many birds, owls are enemies. Some species, such as the diurnal pygmy owl (*Glaucidium gnoma*), are "a terror among songbirds, woodpeckers, and even quail" (Tyler and Phillips, 1978). The screech owl (*Otus asio*) and the flammulated owl (*Otus flammeolus*) also commonly feed on birds, while the elf owl (*Microthene whitneyi*) is almost exclusively insectivorous and, according to Tyler and Phillips (1978), "never takes birds of any kind." The harmless elf owl does not look radically different from the bird-eating pygmy owl, so it might be better for a bird to assume that most owls are dangerous, rather than consider itself to be an excellent owl taxonomist and pick out those few who are not. And then, the same owl species do not always eat the same things. The barn owl (*Tyto alba*) eats only rodents in some areas, but its pellets show bird remains in others (Tyler and Phillips, 1978). In short, if it looks like an owl, then there is a good chance that it eats birds.

Whether or not a *mobbing* bird places itself at great risk is another matter. I only know of one recorded instance of avian mobbers actually being attacked by an owl. Denson (1979) saw a great horned owl "quickly extending its left foot" and capture a bold mobbing crow who got too close. But even this fatality may have been due to a misjudgment by the victim; crows do not need to get *that* close in order to mob.

The question of why a crow or some other animal might mob a predator was, and continues to be, of considerable interest to evolutionary biologists, following Darwin's challenge that his theory of evolution by natural selection would suffer a fatal blow if it could be shown that one animal's actions were solely directed for the benefit of another's. However, following the insight of W. D. Hamilton (1964) that an individual's total fitness depends both on the genes passed on through one's offspring (individual fitness) as well as on the copies of one's genes passed on through other relatives (kin selection), J. Maynard Smith (1965) suggested that the altruistic behavior of alarm calling might be explained through kin selection. Trivers (1971) went on to suggest that altruism could also evolve by a mechanism independent of kin selection. His model, called "reciprocal altruism," concerns one individual helping another who will later repay the favor. This model might apply in those (usually long-term) associations where individuals are recognized so that cheaters (those who take no risks yet take advantage of the risks taken by others) can be discriminated against. Since many birds are capable of individual recognition, and since many form long-term associations of related as well as unrelated individuals, both the kin selection and reciprocal altruism models could potentially apply to the evolution of mobbing (and the associated alarm calling).

Mobbing might also be a form of direct parental investment where the adults protect their young by (1) distracting the predator; (2) leading it away; or (3) harming it. In addition, a mobber might gain direct personal advantage by advertising to the predator that it has been seen and is capable of escape, so that pursuit is futile (Smythe, 1970). Some of the "mobbers" might only be attending at an owl to learn from the mobbers who their enemy is (Curio and Vieth, 1978).

The many theoretical possibilities of why mobbing might occur has stimulated a wealth of empirical studies. Not surprisingly, it has not generated any universal answer, possibly because there are multiple contexts or multiple answers. In order to give a general overview and provide a broad perspective, I will not restrict myself to a discussion of the mobbing of owls because similar principles will probably apply to mobbing of other enemies as well.

The word "mobbing" implies a group, or many individuals. However, as mentioned earlier, it may have many functions unrelated to group assault on an enemy. Also, mobbing by birds starts with a single individual, and discussion of mobbing behavior restricted to groups may be artificial because so far there is no demonstrable difference in behavior or function whether one or several birds are harassing an owl.

Mobbing birds hop around in the vicinity of their enemy, fly at it, flick their tails and wings, and give alarm calls. In short, they draw attention to themselves by visual cues and by calling behavior that accentuates these cues. In most reviews, "alarm calling" has been singled out as separate from the overall mobbing behavior (Wilson, 1975; Harvey and Greenwood, 1978). Since at least in avian mobbing alarm calling is a significant component of the mobbing response, a discussion of mobbing without considerable attention to alarm calling is incomplete.

Part of the problem of making the separation is in the semantics of "alarm" calling. Not all calls an animal makes near a predator function in mobbing. For example, birds have "distress" screams or calls given when they are captured, and one hypothetical function of these screams is to attract mobbers (Stefanski and Falls, 1972), especially kin (Rohwer, Fretwell, and Tuckfield, 1976). Other calls variously described as distress calls, fear trills, "seet" calls, hawk alarm calls, or simply "alarm calls" have acoustical properties that should make them difficult to localize (Marler, 1957). Such calls are given by birds, who (in contrast to mobbers) hide from the predator; they are thought to be a "take-cover" signal. The giving of *these* "alarm calls" was thought to involve altruism, whereby the caller places itself in danger by drawing attention to itself. But Charnov and Krebs (1975) theorize that even though the callers may help flockmates, the take-cover signals might be given because they benefit the caller directly because they manipulate the flock in scurrying for cover, thereby providing alternate targets for the predator and deflecting attack from the caller. These kinds of alarm calls, then, only make sense if flockmates are present to take flight near the predator.

In contrast, avian "mobbing calls" are not made from hiding, and at least the calls I observed were frequently made by individuals even when there were no other adult birds nearby. Such calls are short sounds that might provide instant fixation of the sound source (Marler, 1957). Nevertheless, there is disagreement about whether or not the acoustical properties of the two kinds of calls are effectively different to predators. Pygmy owls (*Glaucidium perlatum* and *G. brasi-*

lianum) and goshawks (*Accipiter gentilis*) are able to localize the alarm calls despite their apparent ventriloquial qualities (Shalter and Schleidt, 1977; Shalter, 1978).

Mobbing (and associated alarm calling) is almost ubiquitous in passerine birds against owls. And it is directed by many of them against some other passerines (especially crows and their relatives), against some hawks, snakes, and even humans and other mammals. Barn swallows, for example, mob birds of prey, domestic dogs and cats, red squirrels, chipmunks and humans (Shields, 1984). Alarm calling (without mobbing) is also found in some ungulates (Hirth and Mc-Cullough, 1977), primates (Tenaza and Tilson, 1977; Seyfarth, Cheney, and Marler, 1980), as well as in colonial ground squirrels (Owings and Coss, 1977; Hennessey and Owings, 1978); Hennessey et al., 1981; Sherman, 1977, 1981) against mammalian and avian predators (Sherman, 1985).

There is relatively little evidence that callers put themselves at risk. Barash (1975), during seven years of studying free-living marmots saw only eight instances of natural predation, and not one of these was preceded by an alarm call by the victim!

But Sherman (1977, 1985) found that in Belding's ground squirrels (*Spermophilus beldingi*) animals may place themselves in danger when they (primarily the females) alarm call by trilling repeatedly in the presence of mammalian predators (weasals, badgers, coyotes, dogs, and marten); they are sometimes killed and eaten. In contrast, in the presence of aerial predators the squirrels (both sexes) whistle once each, which in contrast directly benefits the callers by facilitating their escape, perhaps through a confusion effect or safety in numbers (Sherman, 1985). Sherman investigated several alternative hypotheses, and since the dangerous trill alarm calling was done primarily by females in the presence of kin, he concluded that it best fitted the kin selection hypothesis. On the other hand, the whistle to hawks was unaffected by kinship. Alarm calling in a crowd of birds mobbing an owl could conceivably have a kin warning function where parents, sisters, brothers, aunts, and uncles of the caller benefit. Unfortunately, to my knowledge this has so far never been quantitatively studied.

Overall, the evidence for *avian* callers or mobbers placing themselves at risk is almost nonexistent, yet most theories of alarm calling and mobbing in birds rest on the assumption of risk. On the other hand, the evidence that the beneficiaries of calling are offspring as well as other related individuals is excellent. However, in most cases it is difficult to differentiate whether the beneficiaries are offspring or other kin (Shields, 1980). On a fundamental theoretical basis it probably makes little difference. All phenotypic altruism is likely genotypic selfishness, or it would not have evolved. In any case, it makes no difference what one calls it, so long as one knows precisely what happens in each case.

Alarm calling and/or mobbing in birds might both easily have the same effect as alarm calling in squirrels—benefiting sons and daughters. Few will mistake the evolutionary function of the squeeking and wing dragging of a grouse mother in the presence of a fox. It is a distraction display, and it is most vehement in precocial birds on hatching of the young, and in altricial birds upon fledging (Armstrong, 1966). The young (or the parents' reproductive investment) are clearly the beneficiaries of the behavior. Birds mobbing a crow or an owl near its nest might also serve to put on a distraction display with the offspring as beneficiaries, whether they do it as a group response by a colonial ("mobbing") or a solitary nesting bird. The results of a recent study (Shields, 1984) of mobbing in semicolonial barn swallows, *Hirundu rustica*, are consistent with the above. Shields (1984) concluded that of the various potential hypotheses considered, the one most consistent with predictions is that the birds' mobbing response serves to defend the young. Although several birds might simultaneously mob an owl or some other predator, there was no evidence that the birds acted to defend anything other than their own young; the group was only a "pseudo-group."

In a colony, however, the mob of many may achieve what a single bird cannot. For example, Kruuk (1964) showed that colonially nesting black-headed gulls mob carrion crows as well as herring gulls (which eat their eggs and chicks) by divebombing on these nest enemies while screaming loudly. In addition, the gulls released apparently judiciously timed (unproven) semiliquid fecal matter during their dives. The crows and gulls can avoid the divebombing mobbers, but only by continually facing them. As a consequence they are distracted enough to be impeded in their search for eggs and young while under attack. Black-headed gulls, meanwhile, incur little bodily risk by their mobbing, because the crows and herring gulls do not fight back. Similarly, Hoogland and Sherman (1976) showed that bank swallow mobs were effective in deferring blue jays from killing fledgelings, whereas individual swallows would not likely be able to deter a blue jay.

Eberhardt Curio and his colleagues (1985) at the Ruhr University in Bochum, West Germany, point out that in mobbing for nest defense, a bird with a high reproductive value should risk less in defense than one of low value. Thus a chickadee with an average lifespan of two years might be expected to vigorously defend its brood because that might be its entire lifetime productive investment, whereas a long-lived sea bird should take fewer chances because it can invest in reproduction in later years.

Curio (1978), who has examined avian mobbing for two decades, and Vieth, Curio, and Ernst (1980) review a number of hypotheses regarding mobbing behavior toward owls. The three primary (not always mutually exclusive) ones suggested are: (1) to physically defend the nest from predators; (2) to distract the predator's attention so that it does not see the nest and/or offspring; and (3) to lure or drive the predator away (from the nest and elsewhere, such as a roost), the so-called "move-on" hypothesis. Douglas H. Shedd (1982, 1985) provided evidence from the move-on hypothesis by showing that *migrating robins* (*Turdus migratorius*) do not mob, presumably because when night comes they are no longer in the vicinity of the owl they may have encountered during the day. Black-capped chickadees, however, who are permanent residents, mob year round, although mobbing intensity peaks in the breeding season. The various functions of mobbing are difficult to separate because a bird may be protecting or defending the nest both by attempting to drive a predator away as well as by distracting it.

Mobbing is often triggered by diverse animals, even within the same species. For example, the pied flycatcher, *Ficedula hypoleuca*, mobs not only owls but also shrikes, cats, and people (Curio, 1959). To find out what visual stimuli elicit mobbing in the pied flycatcher, Curio (1975) performed a detailed set of experiments using model owls and shrikes on which he systematically altered various features. He used both experienced and hand-reared birds who had never seen owls or shrikes. Both groups mobbed these enemies, and Curio, as have others (Altman, 1956; Cully and Ligon, 1976) concluded that recognition of owls and shrikes is not necessarily based on prior experience. Learning by experience, however, is also involved. For example, Ramsey (1950) reared two young crows with a barred owl, and neither species later showed any alarm in the presence of the other. Indeed, the young crows occasionally even begged from the owl. The song sparrow, *Melospiza melodica*, appears to have an inborn pattern-recognition of owls, a fear of cats and hawks because of their fast movements, and an alarm response to cowbirds through conditioning (Nice and Pekwyk, 1941).

Some individual birds do not respond to a stuffed owl, even one near their nest, or their response is brief. Nevertheless, if they are then given a live owl they mob it vigorously, after which they will also mob the stuffed owls (Shalter, 1978). However, in later experiments Curio, Ernst, and Vieth (1978) showed that the European blackbird *Turdus merula* (a close relative of the American robin, *Turdus migratoria*) learns what to mob by the example of other birds. Curio (1978b) suggested that another function of mobbing by some individual birds is enemy recognition, where the bird benefits from others' experience with little cost of personal experience (see also Kruuk, 1964; Frankenberg, 1981). If this is the case, however, it is not clear why the benefitor would participate in a dangerous activity rather than sit by and watch silently.

CONCLUSIONS

My scientific objective in keeping Bubo was to gain insights into why many birds mob owls. I did not start out to test one or another of the various hypotheses. Indeed, I did not know what the various hypotheses were. However, after watching the owl for almost three summers, with an open-ended question in mind, I could not help but form opinions on the topic relative to what has been thought and done.

Although the great horned owl shows a strong preference for capturing and eating rabbits over other kinds of prey, it will also take many birds when they are plentifully available and rabbits are rare. Certainly Bubo ate dead feathered animals with gusto, so if great horned and presumably other owls in the wild generally eat few birds, it is likely because they are having difficulty finding or catching them relative to other prey, and not because they dislike eating them.

If owls do not care too much whether their meat comes clothed in feathers or in fur, then birds may incur some risk of being captured when they expose themselves by mobbing an owl. There is a report of

a great horned owl who reached out with its foot to grab a nearby mobbing crow (Denson, 1979), but there is no evidence that mobber birds
need to, or in fact do, endanger themselves to owls. This may be because not many people have been able to watch the undisturbed hunting of owls. Or is the idea that mobbing birds endanger themselves a
myth?

In my studies I sought to gain, first and foremost, insights into the
risks of capture a mobbing songbird might incur by examining Bubo's
hunting behavior and the details of his reactions to mobbers. Birds are
not an insignificant part of the diet of many owls, as determined by
analysis of pellet remains. However, it seemed to me that one of the
main unknowns (upon which the resolution of the questions of mobbing impinge), is *how* the birds are captured. Do the owls, because of
their superior night vision, pluck their victims from a branch while
asleep? Are only weak young or injured birds taken? Or are *mobbers*
chased and captured? Few owls will allow a human close enough in
the daytime to enable him or her to find out how mobbers are treated.
Even if one gets close, the wild owl might not chase mobbing birds because it fears the human observer. Still fewer owls will allow themselves to be followed by a human on their hunting excursions, even if
it were possible at night. Pellets only tell you what the owl ate, not
how it hunted.

Bubo was not intimidated by me, so that if there was anything
catchable within sight that he was interested in, he paid no overt attention to me. I was therefore able to observe his hunting behavior unobtrusively. This behavior was notable for what he could and would
do, as well as for what he apparently could not or would not do. His
urge to attack could be triggered even when he was not hungry, and
when triggered it was energetic and persistent. With little or no apparent prior coaching, he attacked crows only after watching them intently as if to determine whether or not they were watching him.
When they were watching him or looked at him, he did not attack, nor
did he continue an attack when his potential victim, such as a crow,
looked up, indicating that it had seen him approaching. Nevertheless,
he vigorously pursued fully feathered young blue jays and white-
throated sparrows that were aware of him and trying to escape, although their flight was clumsy. Without obvious prior experience he
instantly attacked the very first injured squirrel that he saw, whereas
healthy noisy ones were only watched and never attacked. Although I

often saw birds noisily mobbing directly in front of him (within one-half to one meter) I did not see him make a single attempt to capture any of them. It was not lack of hunger that kept him from trying to catch these birds because he spent several days trying to catch (non-mobbing) crows even after he was fully fed. In summary, what Bubo apparently could *not* do, and did not even attempt to do, was to capture a healthy adult bird or squirrel that was aware of him and let him know this by its display to him. He seemed to be acutely aware of whether or not he had the advantage in an attack, and that advantage, against healthy adult birds, might be only at night. An owl's hearing and vision are keener than that of most other birds, enabling them to strike in semidarkness—and some can even do it in total darkness. Their nocturnal and crepuscular habits should give them an advantage for surprise attack against diurnal prey, and their large wings and soft feathers permit silent flight. My observations with Bubo indicate that at least he was not suited to pursue fleeing prey like an accipitrine hawk is.

Of course, most predators take advantage of surprise, when they can manage it. But to underscore the idea that owls are specialists in surprise and not in pursuit, owls such as Bubo, as I have indicated in the text, sometimes molt their retrices, or tail feathers, relatively synchronously, rather than regrowing individual feathers before shedding other ones. Except for the diurnal ferruginous, pygmy, and hawk owls, most nocturnal owls have relatively short tails compared with hawks. Bird-hunting hawks who pursue their prey in forests use their long tails as rudders for rapid maneuvering in flight. I speculate that the nocturnal forest owls are able to get by with a short tail, and sometimes with no tail, because when they rely on surprise attacks, they do not need to be able to maneuver rapidly.

The owl's structure, molt pattern, keen senses, nocturnal hunting, its studious appraisal of vulnerability in prey, and its clumsiness or its unwillingness to pursue suggest that an owl is a specialist as a stealth predator. Therefore a mobbing bird's risk of capture is very slight. In other words, mobbing an owl—at least a nocturnal one like Bubo—is most likely *not* a dangerous, and therefore not an altruistic, act. On the contrary, Bubo, for example, attacked *only* nonmobbing birds. Nevertheless, there are presumably advantages in mobbing, simply because birds would not otherwise evolve to spend large amounts of time and energy in useless activity.

Mobbing may indeed advertise to a potential predator that pursuit would be futile (Smythe, 1970). However, I doubt very much that the evolution of mobbing can be explained by this hypothesis for two reasons. First, the mobbers would not need to be so persistent in their activity (some individuals mobbed more than an hour at a time) nor come so close (they would repeatedly fly within centimeters of Bubo) in order merely to let him know that they have seen him and are alert.

Because there are almost no physical risks involved in mobbing an owl, the major potential cost that has usually been assumed is almost nonexistent, and mobbing intensity against owls is then probably more directly a function of advantages. What are these advantages? From my observations of leading Bubo through the woods it became obvious that mobbing vigor was in most cases almost strictly related to his nearness to young (seldom to eggs), either in or out of the nest, whereas crows and jays, who steal eggs, *are* mobbed near nests with eggs. When Bubo came near the young, the parents mobbed him, and they did not stop until he had left the area. These results are in accord with many other studies on owl mobbing (Altmann, 1956; Curio, 1978b; Shedd, 1982), supporting the conclusion that mobbing an owl is a form of parental care that is functionally similar to a distraction display, like the broken-wing display of many ground-nesting birds who also make themselves conspicuous to a potentially dangerous predator.

The precise nature of how mobbing birds protect their offspring is not quite clear. Curio (1978b) suggests that the mobbers cause the owl to leave the area (the "move-on" hypothesis), or to distract it so that it does not find the young. Since most of the early classical studies of mobbing have used stuffed owls, neither hypothesis has been adequately tested. After watching Bubo, however, I believe that both apply—Bubo both fled from and was distracted by very vigorous mobbing. Both the blue jays and hermit thrushes kept flying at the back of his head, which seemed to irritate him so that he continually turned his head and followed them with his eyes, thus preventing an attack from the bird being watched (though not its mate). Since both members of a pair joined in the alarm-calling, mobbing, response, Bubo was kept almost continually occupied by intently watching one or the other attacker. After Bubo left the area, he was immediately left alone.

Curio's "move-on" hypothesis (see also Bildstein, 1982) is also supported by the mobbing reactions of nonbreeding birds. Like Shedd (1982), I found that only resident birds (black-capped chickadees, blue jays, crows) mobbed him *outside* their breeding season. However, migrating warblers, vireos, and flycatchers, who would probably be in some other area by nightfall, did *not* mob him; I routinely observed dozens of them as they foraged near him without mobbing him or showing any alarm, even when they were in the same tree with him. (These species are suitable food—Bubo ate dead ones routinely.) Once an owl's success in catching a healthy bird depends on surprising it, possibly at night, there should be an advantage for resident territory-holding birds to get the owl to leave the area before nightfall. Since crows have conspicuous roosts to which they return each night, the "move-on" hypothesis should apply especially strongly to them. And indeed, the vigor of the crows' mobbing in winter is surpassed by few other birds, even in spring.

If the idea is correct that resident nonbreeding birds have evolved to mob an owl because they gain the benefit of a safer night at little risk, then one can predict that they would not mob a predator that hunts in the daytime and could be a big danger to them, unless they have help from others and/or unless there are very high benefits. This prediction has some support from at least a few anecdotal observations, but it has not been tested. For example, during the winter in Vermont many small birds, especially chickadees and nuthatches, come to the bird-feeders by our house. On several occasions I have seen northern shrikes, *Lanius excubitor*, in the vicinity. These birds are reputed to be predators of small birds, and from what I have seen this reputation is amply deserved. These shrikes swoop up to a perch, and their heads turn rapidly as they search—it is not hard to guess for what. On one occasion a shrike chased a chickadee from the birdfeeder. The chickadee fled through the woods, and then as it tried to cross a field the shrike pounced on it, knocked it to the ground, killed it, and ate it. On another occasion I saw a shrike chase a white-breasted nuthatch around and around through the trees near the house. The nuthatch finally saved itself by literally diving into a bird box with an entrance hole too small to admit the shrike. In neither case did any of the dozens of birds near the feeders mob; they all remained silent and left hastily. Clearly these small birds are in danger from this daytime-hunting predator. Although birds mob shrikes near their nests in the summer

(Curio, 1978b; Cade, 1962) when the benefits are great (protection of young), perhaps they do not do so in the winter because it is very risky, and with no young to defend, there are few benefits.

Many birds act as if they are aware that different kinds of predators impose different risks and/or that the same predator poses different risks under different conditions. For example, rather than mob a dangerous hawk, some birds stay in hiding and give high-pitched "hawk-call" alarm notes that are difficult to locate (Marler, 1957). Under some conditions, however, the same hawks that are otherwise avoided are also mobbed. Bildstein (1982) showed that marsh harriers, *Circus cyaneus*, elicited six times more mobbing reactions from birds when they were carrying something (voles, or nesting material) than when they were not. He suggested for an explanation that those predators carrying something posed a reduced threat to potential mobbers. Similarly, ground squirrels from populations exposed to venomous snakes show weak mobbing response to snakes compared with those from areas where there are no venomous snakes (Owings and Coos, 1977). Vervet monkeys, on the other hand, give the most alarm calls to predators to which they themselves, rather than their offspring, were most vulnerable (Cheney and Seyfarth, 1981).

Perhaps owls in general are vigorously mobbed because this is a low risk/high benefit behavior. If so, then for a given potential gain, diurnal owls that sometimes pursue birds should be mobbed less vigorously than nocturnal owls. There may be great benefits in mobbing bird-catching hawks and shrikes, but in general the risks may outweigh these benefits. Most likely, dangerous birds are mobbed primarily by colonial birds, where the risks of an individual's mobbing behavior is less, the chances of success are greater, and the presence of kin as beneficiaries provides an added bonus.

In my opinion much of the early literature suffers from making assumptions that are too simple. Birds may be much more sophisticated than we assume. Kruuk (1964) already showed that gulls are less aggressive toward a person carrying a stick than to one who is unarmed, presumably because they recognize the increased risk. More recent studies show that magpies (Buitron, 1983) and red-winged blackbirds (Knight and Temple, 1986) recognize *individual* humans, and are more aggressive toward those they have met (and successfully "repelled") before. Indeed, red-winged blackbirds are more aggressive to a mounted (dead) crow near their nest than to a live one (Knight and

Temple, 1986), again suggesting they scale their attack relative to the perceived risk versus the perceived benefit; they become bolder after they have been successful once.

A single unifying theory to help explain mobbing is unlikely. Perhaps the only generalization that will emerge is that mobbing will be expressed in varying ways and degrees depending on the benefit/cost ratios involved with different combinations of potentiel prey and predators. Mobbing of owls by songbirds in the daytime is apparently at the high benefit/low risk end of the spectrum because of the owls' role as stealth predators. At present, the real unknown is not so much the evolutionary significance of mobbing, but what goes on in the brain of the mobbers and the mobbed.

REFERENCES

Adamcik, R. S., and L. B. Keith. 1978. Regional movements and mortality of Great Horned Owls in relation to snowshoe hare fluctuations. *Can. Field Nat.* 92(3): 228–234.

Adamcik, R. S., A. W. Todd, and L. B. Keith. 1978. Demographic and dietary responses of Great Horned Owls during a snowshoe hare cycle. *Can. Field Nat.* 92: 156–166.

Altmann, S. A. 1956. Avian mobbing behavior and predator recognition. *Condor* 58: 241–253.

Andersson, M. 1982. Reproductive tactics of the long-tailed skua *Stercorarius longicaudus*. *Oikos* 37(3): 287–294.

Andersson, M., and R. A. Norberg. 1981. Evolution of reversed sexual size dimorphism and role partitioning among predatory birds, with a size scaling of flight performance. *Biol. J. Linn. Soc.* 15: 105–130.

Angell, Tony. 1974. *Owls*. Seattle: University of Washington Press.

Armstrong, E. A. 1966. Distraction display and the human predator. *Ibis* 98: 641–654.

Austin, Oliver L. 1932. An interesting great horned owl capture. *Bird Banding* 3: 33.

Austing, G. Ronald, and John B. Holt, Jr. 1966. *The World of the Great Horned Owl*. Philadelphia and New York: Lippincott.

Barash, D. P. 1975. Marmot alarm-calling and the question of altruistic behavior. *Amer. Midl. Natur.* 94: 468–470.

Bartholomew, G. A., R. C. Lasienski, and E. C. Crawford. 1968. Patterns of panting and gular flutter in cormorants, pelicans, owls, and doves. *Condor* 70: 31–34.

Baumgarten, Frederick M. 1939. Territoriality and population in the Great Horned owl. *Auk* 56: 274–282.

Bendire, Charles F. 1892. Life histories of North American birds. *U.S. Nat. Mus. Spec. Bull.* 1: 325–414.

Bent, Arthur C. 1961. *Life Histories of North American Birds of Prey*, part 2, *Hawks, Falcons, Caracaras and Owls*. New York: Dover.

Biederman, B. M., D. Florence, and C. C. Lin. 1980. Cytogenetic analysis of great horned owls (*Bubo virginianus*) Cytogenet. *Cell Genet.* 28(1/2): 79–86.

Bildstein, K. L. 1982. Responses of Northern Harriers to mobbing passerines. *J. Field Ornithol.* 53: 7–14.

Blest, A. D. 1957. The function of eyespot patterns in the Lepidoptera. *Behaviour* 2: 209–256.

Bowmaker, J. K., and G. R. Martin. 1978. Visual pigments and color vision in a nocturnal bird, *Strix aluco* (Tawny owl). *Vision Res.* 18(9): 1125–1130.

Bralliar, Floyd. 1922. As quoted by A. C. Bent, 1961.

Brockway, A. W. 1918. Large flight of Great Horned Owls and Goshawks at Hadlyme, Connecticut. *Auk* 35: 351–352.

Buitron, D. 1983. Variability in the responses of black-billed magpies to natural predators. *Behaviour* 88: 209–235.

Burton, John A. 1973. *Owls of the World: Their Evolution, Structure and Ecology*. New York: E. P. Dutton.

Bunn, D. S., A. B. Warburton, and R.D.S. Wilson. 1982. *The Barn Owl*. Vermillion, S.D.: Buteo Books.

Cade, T. J. 1962. Wing movements, hunting and displays of the Northern Shrike. *Wilson Bull.* 74: 386–408.

Cade, T. J., and Y. Greenwald. 1966. Nasal salt secretion in falconiform birds. *Condor* 68: 338–350.

Cameron, A., and P. Parnall. 1981. *The Nightwatchers*. New York: Four Winds Press.

Cawthorn, R. J., and P.H.G. Stockdale. 1981. Description of *Eimeria bubonis*, new species (Protozoa: Eimeridae) and *Caryospora bubonis*, new species (Protozoa: Eimeridae) in the great horned owl, *Bubo virginianus*, of Saskatchewan, Canada. *Can. J. Zool.* 59(2): 170–173.

Chaplin, S. B., D. A. Diesel, and J. A. Kasparie. 1984. Body temperature regulation in Red-Tailed Hawks and Great Horned Owls: Responses to air temperature and food deprivation. *Condor* 86: 175–181.

Charnov, E. L., and J. R. Krebs. 1975. The evolution of alarm calls: Altruism or manipulation? *Am. Nat.* 109: 107–112.

Cheney, D. L., and R. M. Seyfarth. 1981. Selective forces affecting the predator alarm calls of vervet monkeys. *Behaviour* 76: 25–61.

Clark, Richard J., Dwight G. Smith, and Leon H. Kelso. 1978. *Working Bibliography of Owls of the World*. National Wildlife Federation.

Craighead, John J., and Frank C. Craighead, Jr. 1969. *Hawks, Owls, and Wildlife*. New York: Dover.

Cully, J. F., Jr., and J. D. Ligon. 1976. Comparative mobbing behavior of Scrub and Mexican jays. *Auk* 93: 116–125.

Curio, E. 1959. Verhaltensstudien am Trauerschnäpper. *Z. Tierpsychol.* 3: 1–118.

Curio, E. 1975. The functional organization of anti-predator behavior in the pied flycatcher: A study of avian perception. *Anim. Behav.* 23: 1–115.

Curio, E. 1978a. Cultural transmission of enemy recognition: One function of mobbing. *Science* 202: 899–901.

Curio, E. 1978b. The adaptive significance of avian mobbing. I. Teleonomic hypotheses and predictions. *Z. Tierpsychol.* 48: 175–183.

Curio, E., U. Ernst, and W. Vieth. 1978. The adaptive significance of avian mobbing. II. Cultural transmission of enemy recognition in blackbirds: Effectiveness and some constraints. *Z. Tierpsychol.* 48: 184–202.

Curio, E., K. Regelmann, and U. Zimmermann. 1985. Brood defence in the great tit (*Parus major*): The influence of life-history and habitat. *Behav. Ecol. Sociobiol.* 16: 273–283.

Delmee, E., P. Dachy, and P. Simon. 1981. A tawny owl (*Strix duco*) close to 19 years old (1960–1979). *Gerfaut* 70(2): 201–210.

Denson, R. D. 1979. Owl predation on a mobbing crow. *Wilson Bull.* 91: 133.

Dice, Lee R. 1945. Minimum intensities of illumination under which owls can find dead prey by sight. *Amer. Natur.* 79(784): 385–413.

Dice, Lee R. 1947. Effectiveness of selection by owls of deer-mice (*Peromyscus maniculatus*) which contrast in color with their background. *Contr. Lab. Vertebr. Biol. Univ. Mich.* 34: 1–20.

Dodsen, P., and D. Wexler. 1979. Taphonomic investigation of owl pellets. *Paleobiology* 5(3): 275–284.

Duke, G. E., O. A. Evanson, and A. Jegers. 1976. Meal to pellet intervals in 14 species of captive raptors. *Comp. Biochem. Physiol. A. Comp. Physiol.* 53(1): 1–6.

Duke, G. E., and D. D. Rhoades. 1977. Factors affecting meal to pellet intervals in great horned owls (*Bubo virginianus*). *Comp. Biochem. Physiol. A. Comp. Physiol.* 56(3): 283–286.

Earhart, M. C., and N. K. Johnson. 1970. Size dimorphism and food habits of North American owls. *Condor* 72(3): 251–264.

Emerson, K. C. 1961. Three new species of Mallophaga from the great horned owl. *Proc. Biol. Soc. Wash.* 74: 187–192.

Emlen, J. T. 1973. Vocal stimulation in the Great Horned Owls. *Condor* 75(1): 126–127.

Erkert, H. G. 1967. Beleuchtungsabhängige Aktivitätsoptima bei Eulen und circadiane Regel. *Naturwiss.* 54(9): 231–232.

Erkert, Hans G. 1969. Die Bedeutung des Lichtsinnes für Aktivität und Raumorientierung der Schleiereule (*Tyto alba guttata* Brehm.). *Z. Vergl. Physiol.* 64(1): 37–70.

Errington, P. L. 1944. To babes really lost in the woods. *Can. Field Natur.* 58(2): 52–54.

Errington, Paul L. 1932. Studies in the behavior of the Great Horned Owl. *Wilson Bull.* 44: 212–220.

Everett, Michael. 1977. *A Natural History of Owls.* London: Hamlyn Publ. Group Limited.

Fite, K. V. 1973. Anatomical and behavioral correlates of visual acuity in the Great Horned Owl. *Vision Res.* 13(2): 219–230.

Fleay, David. 1968. *Nightwatchmen of Bush and Plain (Australian Owls and Owl-like Birds).* Brisbane: The Jacaranda Press.

Forsman, E. D., and H. M. Wright. 1979. Allopreening in owls (*Strix occidentalis*): What are its functions? *Auk* 96(3): 525–531.

Frankenberg, F. 1981. Adaptive significance of avian mobbing: 4. Alerting others and perception advertisement in blackbirds (*Turdus merula*) facing an owl (*Athene noctua*). *Z. Tierpsychol.* 55(2): 97–118.

Fujioka, Masahiro. 1985. Sibling competition and siblicide in a synchronously-hatched brood of the cattle egret *Bubulcus ibis. Anim. Behav.* 33: 1228–1242.

Fuller, M. R., and G. E. Duke. 1979. Regulating pellet egestion: The effect of multiple feeding on meal to pellet intervals in great horned owls. *Comp. Biochem. Physiol. A. Comp. Physiol.* 62(2): 439–444.

Griffee, W. E. 1958. More mice—more owl eggs. *Murrelet* 39(2): 18.

Grimes, Samuel A. 1936. Great horned owl and common black snake in mortal combat. *Florida Nat.* 9: 77–78.

Grimm, Robert J., and W. M. Whitehouse. 1963. Pellet formation in a Great Horned Owl: A roentgenographic study. *Auk.* 80(3): 301–306.

Hagar, Donald C., Jr. 1957. Nesting populations of red-tailed hawks and horned owls in central New York State. *Wilson Bull.* 69(3): 263–272.

Hamerstrom, F. 1956. The influence of a hawk's appetite on mobbing. *Condor* 59: 192–194.

Hamilton, W. D. 1964. The evolution of social behavior. *J. Theor. Biol.* 7: 1–52.

Harvey, P. H., and P. J. Greenwood. 1978. Antipredator defense strategies: Some evolutionary problems. In *Behavioral Ecology*, J. R. Krebs and N. B. Davies, pp. 129–151. Sunderland, Mass.: Sinauer Associates.

Hecht, Selig, and M. H. Pirenne. 1940. The sensibility of the nocturnal long-eared owl in the spectrum. *J. Gen. Physiol.* 23(6): 706–717.

Heinrich, B. 1979. *Bumblebee Economics*. Cambridge, Mass.: Harvard University Press.

Hennessy, D. F., and D. H. Owings. 1978. Snake species discrimination and the role of olfactory cues in the snake-elicited behavior of the California ground squirrel. *Behaviour* 65: 115–124.

Hennessy, D. F., D. H. Owings, M. P. Rowe, R. B. Coss, and D. W. Leger. 1981. The information afforded by a variable signal: Constraints on snake-elicited tail flagging by California ground squirrels. *Behaviour* 78: 188–226.

Herms, W. B., and C. G. Kadner. 1937. The louse fly, *Lynchia fusca*, parasite of the owl, *Bubo virginianus pacificus*, a new vector of malaria of the California Valley Quail. *J. Parasitol.* 23(3): 296–297.

Hirth, D. H., and D. R. McCullough. 1977. Evolution, alarm signals in ungulates with special reference to white-tailed deer. *Amer. Natur.* 111: 31–42.

Hocking, B., and B. L. Mitchell. 1961. Owl vision. *Ibis* 103(2): 248–288.

Holland, H. M. 1926. Who would have thought it of *Bubo? Bird-Lore* 38: 1–4.

Hoogland, J. L., and R. W. Sherman. 1976. Advantages and disadvantages of bank swallow (*Riparia riparia*) coloniality. *Ecol. Mon.* 46: 33–58.

Houston, C. S. 1978. Recovery of Saskatchewan-banded great horned owls. *Can. Field. Nat.* 92(1): 61–66.

Ingram, C. 1959. The importance of juvenile cannibalism in the breeding biology of certain birds of prey. *Auk.* 76: 218–226.

Jackson, R. W. 1925. Strange behavior of Great Horned Owl in behalf of young. *Auk* 42: 445.

Karalus, Karl E., and Allan W. Eckert. 1974. *The Owls of North America.* Garden City, N.Y.: Doubleday.

Kaufman, D. W. 1974. Adaptive coloration in *Peromyscus polionotus*: Experimental selection by owls. *J. Mammal.* 55(2): 271–283.

Kelso, Leon, and E. H. Kelso. 1936. The relation of feathering of feet of American owls to humidity of environment and to life zones. *Auk.* 53(1): 51–56.

Keyes, Charles R. 1911. A history of certain Great Horned Owls. *Condor* 13: 5–19.

Knight, R. L. 1984. Responses of nesting ravens to people in areas of different human densities. *Condor* 86: 345–346.

Knight, R. L., and Stanley A. Temple. 1986. Methodologized problems in studies of avian nest defence. *Animal Behaviour* 34: 561–566.

Knudsen, Eric I. 1981. The hearing of the barn owl. *Sci. Am.* 245: 83–91.

Knudsen, E. I., and P. F. Knudsen. 1985. Vision guides to adjustment of auditory localization in young Barn owls. *Science* 230: 545–548.

Konishi, M., and A. S. Kenuk. 1975. Discrimination of noise spectra by memory in the barn owl. *J. Comp. Physiol. A. Sens. Neural Behav. Physiol.* 97(1): 55–58.

Kostuch, T. E., and G. E. Duke. 1975. Gastric motility in great horned owls (*Bubo virginianus*). *Comp. Biochem. Physiol. A. Comp. Physiol.* 51(1): 201–206.

Krishan, A., G. J. Haiden, and R. N. Shoffner. 1965. *Chromosoma* 17(3): 258–263.

Kruuk, H. 1964. Predators and anti-predator behaviour of the black-headed gull (*Larus ridibundus* L.). *Behav. Suppl.* 11: 1–129.

Lee, R. D., and R. E. Ryckman. 1954. Coleoptera and Diptera reared from owl nests. *Bull. Brooklyn Ent. Soc.* 49(1): 23–24.

McInvaille, W. B., Jr., and L. B. Keith. 1974. Predator-prey relations and breeding biology of the Great Horned Owl and Red-tailed Hawk in Central Alberta. *Can. Field. Nat.* 88(1): 1–20.

Marler, P. 1957. Specific distinctiveness in the communication signals of birds. *Behaviour* 11: 13–39.

Marti, C. D. 1973. Food consumption and pellet formation rates in four owl species. *Wilson Bull.* 85(2): 178–181.

Martin, G. R., 1974. Color vision in the Tawny Owl (*Strix aluco*). *J. Comp. & Physiol. Psychol.* 86: 133–141.

Martin, G. R., 1977. Absolute visual threshold and scotopic spectral sensitivity in the tawny owl *Strix aluco. Nature* 268: 636–638.

Martin, G. 1978. Through an owl's eye. *New Scientist,* 12 Jan. 1978: 72–74.

Martin, G. R. 1982. An owl's eye: Schematic optics and visual performance in *Strix aluco. J. Comp. Physiol. A. Sens. Neural Behav. Physiol.* 145(3): 341–450.

Maslow, Jonathan E. 1983. *The Owl Papers*. New York: Dutton.

Maynard Smith, J. 1965. The evolution of alarm calls. *Amer. Natur.* 94: 59–63.

Mayr, E., and M. Mayr. 1954. The tail molt of small owls. *Auk* 71: 172–178.

Mendall, H. L. 1944. Food habits of hawks and owls in Maine. *J. Wildlife Mgt.* 8(3): 198–208.

Mikkola, Heimo. 1983. *Owls of Europe*. Vermillion, S.D.: Buteo Books.

Miller, L. 1930. The territorial concept in the horned owl. *Condor* 32: 290–291.

Mowat, Farley. 1961. *Owls in the Family*. Toronto: Little, Brown & Co.

Murphy, C. J., and H. C. Howland. 1983. Owls eyes: Accommodation, corneal curvature and refractive state. *J. Comp. Physiol. A. Sens. Neural Behav. Physiol.* 151(3): 277–284.

Mysterud, I., and H. Dunker. 1979. Mammal ear mimicry: A hypothesis on the behavioral function of owl "horns." *Anim. Behav.* 27: 315.

National Wildlife. June-July 1984, p. 25.

Nero, Robert W. 1980. *The Great Gray Owl: Phantom of the Northern Forest*. Washington, D.C.: Smithsonian Institution Press.

Nero, Robert W. 1984. *Redwings*. Washington, D.C.: Smithsonian Institution Press.

Nero, Robert W. 1985. Gray'l attracts lots of attention. *Man. Naturalists Bull.* 8 (10): 6.

Nice, Margaret M. 1934. A hawk census from Arizona to Massachusetts. *Wilson Bull.* 46(2): 93–95.

Nice, Margaret M. 1937. Studies in the life history of the Song Sparrow. *J. Trans. Linn. Soc. New York* 4: 247.

Nice, M. M., and J. T. Pekwyk. 1941. Enemy recognition in the song sparrow. *Auk* 58: 195–213.

Nicholson, Donald J. 1926. Horned owl shrewdness and ferocity. *Oologist* 43: 14.

Norberg, R. A. 1970. Hunting technique of Tengmalm's Owl *Aegolius funereus* (L.). *Ornis Scandinavica* 1: 51–64.

Norberg, R. A. 1978. Skull asymmetry, ear structure and function, and auditory localization in Tengmalm's Owl, *Aegolius funereus* (L.). *Phil. Trans. R. Soc. Lond.* B 282: 325–410.

Norton, A. H. 1982. Watching a pair of great horned owls. *The Maine Naturalist* 8(1): 3–16.

Oliphant, L. W. 1981. Crystalline pteridines in the stromal pigment cells of the iris of the Great Horned Owl (*Bubo virginianus*). *Cell Tissue Res.* 217(2): 387–395.

Owings, D. H., and R. G. Coss. 1977. Snake mobbing by California ground squirrels: Adaptive variation and ontogeny. *Behaviour* 62: 50–69.

Packard, Robert L. 1954. Great Horned Owl attacking squirrel nests. *Wilson Bull.* 66: 272.

Palmgren, P. 1982. Sound location in owls. *Ornis Fenn.* 59(1): 32–35.

Payne, R. S. 1962. How the barn owl locates prey by hearing. *The Living Bird* 1: 151–189.

Payne, R. S. 1971. Acoustic location of prey by barn owls (*Tyto alba*). *J. Exp. Biol.* 56: 535–573.

Perrone, Michael. 1981. Adaptive significance of ear tufts in owls. *Condor* 83: 383–384.

Philips, J. R., and R. A. Norton. 1978. *Bubophilus ascalaphus* gen. et. sp. n. (*Acaria: Syringophilidae*) from the quills of a great horned owl (*Bubo virginianus*). *J. Parasitol.* 64(5): 900–904.

Ramalingam, S., and W. M. Samuels. 1978. Helminths in the great horned owl, *Bubo virginianus*, and snowy owl, *Nyctea scandiaca*, of Alberta. *Can. J. Zool.* 56(1): 2454–2456.

Ramsey, A. O. 1950. Conditional responses in crows. *Auk* 67: 456–459.

Rausch, R. 1948. Observations on cestods in North American owls. *Am. Midl. Natur.* 40(2): 462–471.

Reed, C. I., and B. P. Reed. 1928. The mechanism of pellet formation in the great horned owl (*Bubo virginianus*). *Science* 68(1763): 359–360.

Rhoades, D. D., and G. E. Duke. 1977. Cineradiographic studies of gastric motility in the Great Horned Owl (*Bubo virginianus*). *Condor* 79(3): 328–334.

Rohwer, S., S. D. Fretwell, and R. C. Tuckfield. 1976. Distress screams as a measure of kinship in birds. *Am. Midl. Natur.* 96: 418–430.

Ryckman, R. E. 1953. Diptera reared from barn owl nests. *Pan-Pacific Ent.* 29(1): 60.

Schlenoff, D. H. 1985. The startle responses of blue jays to *Catocola* (Lepidoptera: Noctuidae) prey models. *Anim. Behav.* 33: 1057–1067.

Schlitter, D. A. 1973. A new species of gerbil from South West Africa with remarks on *Gerbillus tytonis* Bauer and Neithamer, 1959 (Rodentia: Gerbillinae). *Bull. South Calif. Adac. Sci.* 72(1): 13–18.

Service, William. 1969. *Owl.* New York: Knopf.

Seton, Ernest Thompson. 1890. The birds of Manitoba. *Proc. U.S. Nat. Mus.* 13: 457–643.

Seyfarth, R. M., D. L. Cheney, and P. Marler. 1980. Monkey responses to three different alarm calls: Evidence of predator classification and semantic communication. *Science* 210: 801–803.

Shalter, M. D. 1978. Mobbing in the pied flycatcher: Effect of experiencing a live owl on response to a stuffed facsimile. *Z. Tierpsychol.* 47(2): 173–179.

Shalter, M. D., and W. M. Schleidt. 1977. The ability of barn owls *Tyto alba* to discriminate and localize alarm calls. *Ibis* 119: 22–27.

Shedd, D. H. 1982. Seasonal variation and function of mobbing and related antipredator behaviour of the American robin (*Turdus migratorius*). *Auk* 99: 342–346.

Shedd, D. H. 1985. A propensity to mob. *Living Bird* 4: 8–11.

Sherman, P. W. 1977. Nepotism and the evolution of alarm calls. *Science* 197: 1246–1253.

Sherman, P. W. 1981. Kinship, demography and Belding's ground squirrel nepotism. *Behav. Ecol. & Sociobiol.* 8: 251–259.

Sherman, P. W. 1985. Alarm calls of Belding's ground squirrels to aerial predators: Nepotism or self-preservation? *Behav. Biol. Sociobiol.* 17: 313–323.

Shields, W. M. 1984. Barn swallow mobbing: Self defence, collateral kin defence, group defence, or parental care? *Animal Behaviour* 32: 132–148.

Siegfried, W. R., R. L. Abraham, and V. B. Kuechle. 1976. Daily temperature cycles in barred, great horned and snowy owls. *Condor* 77(4): 502–506.

Smith, J.N.M., P. Arcese, and I. G. McLean. 1984. Age, experience, and enemy recognition by wild song sparrows. *Behav. Ecol. Sociobiol.* 14: 101–106.

Smythe, N. 1970. On the existence of "pursuit invitation" signals in mammals. *Am. Natur.* 104: 491–494.

Soper, J. D. 1918. Flight of horned owls in Canada. *Auk* 35: 478.

Sparks, John, and Tony Soper. 1970. *Owls, Their Natural and Unnatural History*. New York: Taplinger.

Speirs, J. M. 1961. Courtship of great horned owls. *Can. Field Natur.* 175(1): 52.

Springer, M. A., and J. S. Kirkley. 1978. Inter and intraspecific interactions between red-tailed hawks and great horned owls in central Ohio. *Ohio J. Sci.* 78(6): 323–328.

Steadman, D. W., G. K. Pregill, and S. L. Olsen. 1984. Fossil vertebrates from Antigua, Lesser Antilles: Evidence for late Holocene human-caused extinctions in the West Indies. *Proc. Natl. Acad. Sci. USA* 81: 4448–4451.

Stefanski, R. A., and J. B. Falls. 1972. A study of distress calls of song, swamp, and white-throated sparrows (Aves: Fringillidae). *Can. J. Zool.* 50: 1501–1525.

Steinbach, M. J. 1973. Eye movements of the owl. *Vision Res.* 13(4): 889–891.

Steinman, M. J., R. G. Angus, and K. E. Money. 1974. Torsion eye movements of the Great Horned Owl. *Vision Res.* 14(8): 745–746.

Stewart, P. A. 1969. Movements, population fluctuations, and mortality among great horned owls. *Wilson Bull.* 81(2): 155–162.

Stresemann, E. 1967. Inheritance and adaptation in molt. Pp. 75–80 in D. W. Snow (ed.), *Proceedings of the XIV International Ornithological Congress*. Oxford and Edinburgh: Blackwell.

Sutton, George Miksch. 1929. How can a bird-lover help to save the hawks and owls? *Auk* 46(2): 190–195.

Tenaza, R. R., and R. L. Tilson. 1977. Evolution of long-distance alarm calls in Kloss's gibbon. *Nature* 268: 891–902.

Thoreau, Henry David. 1972. *The Maine Woods*. Princeton, N.J.: Princeton University Press.

Tietjen, E. 1985. The survivors still sing: The ecology and dynamics of predation. *Habitat* 2: 12–17.

Tinbergen, N., G. J. Brockhuysen, F. Feekes, J.C.W. Houghton, H. Krouk, and E. Szulc. 1962. Egg shell removal by the black-headed gull, *Larus ridibundus* L.: A behaviour component of camouflage. *Behaviour* 19: 76–116.

Tirrell, P. B. 1978. *Protocalliphora avium* (Diptera) myasis in great horned owls (*Bubo virginianus*), red-tailed hawks (*Buteo jamaicensis*), and Swainson's hawks (*Buteo swainsoni*) in North Dakota, USA. *Raptor Res.* 12(1/2): 21–27.

Traylor, M. A. 1958. Variation in South American Great Horned Owls. *Auk* 75(2): 143–149.

Trivers, R. L. 1971. The evolution of reciprocal altruism. *Q. Rev. Biol.* 46: 35-57.

Turner, J. C., Jr., and L. McClanahan, Jr. 1981. Physiogenesis of endothermy and its relation to growth in the Great Horned Owl. *Comp. Biochem. Physiol.* 68A: 167–173.

Tyler, Hamilton, and Don Phillips. 1978. *Owls by Day and Night*. Happy Camp, Calif.: Naturegraph.

Ulrich, S. R. 1984. Views through a window may influence recovery from surgery. *Science* 224: 420–421.

Vanderplank, F. L. 1934. The effect of intra-red waves on tawny owls (*Strix aluco*). *Proc. Zool. Soc. London* 1934(3): 505–507.

Vieth, W., E. Curio, and U. Ernst. 1980. The adaptive significance of avian mobbing. III. Cultural transmission of enemy recognition in blackbirds: Cross species tutoring and properties of learning. *Anim. Behav.* 28: 1217–1229.

Walker, L. W. 1974. *The Book of Owls*. New York: Alfred A. Knopf.

Webster, J. D., and R. T. Orr. 1958. Variation in the Great Horned Owl in Middle America. *Auk* 75(2): 134–142.

Welty, Joel C. 1962. *The Life of Birds*. Philadelphia and London: W. B. Saunders.

Wilkinson, A. Norman. 1913. Horned owl killing a skunk. *Bird-Lore* 15: 369.

Wilson, E. O. 1975. *Sociobiology: The New Synthesis*. Cambridge, Mass.: Belknap Press.

Wilson, E. O. 1985. Time to revive systematics. *Science* 230: 1227.

Woods, J. G. 1971. Insects parasitic on the saw-wheat owl, *Aegulius acadicus* (Gmelin) (Aves: Strigidae). *Can. J. Zool.* 49(6): 959–960.

INDEX

Library of Congress Cataloging-in-Publication Data

Heinrich, Bernd, 1940–
  One man's owl.

  Bibliography: p.
  Includes index.
  1. Horned owl. 2. Wildlife rescue. I. Title.
QL696.S83H45 1987   598'.97   87–3758
ISBN 0–691–08470–X (alk. paper)